Information Theoretic Learning-Based Filter

Haiquan Zhao · Xinyan Hou · Xiaoqiang Long

Information Theoretic Learning-Based Filter

Algorithms, Analysis and Applications

Springer

Haiquan Zhao
Department of Electrical Engineering
Southwest Jiaotong University
Chengdu, China

Xinyan Hou
School of Automation and Electrical Engineering
Lanzhou Jiaotong University
Lanzhou, Gansu, China

Xiaoqiang Long
College of Physics and Electronic Engineering
Northwest Normal University
Lanzhou, Gansu, China

ISBN 978-3-032-29622-1 ISBN 978-3-032-29623-8 (eBook)
https://doi.org/10.1007/978-3-032-29623-8

This Springer imprint is published by the registered company Springer Nature Switzerland AG
The registered company address is: Gewerbestrasse 11, 6330 Cham, Switzerland

Preface

Information Theoretic Learning (ITL) has become an important framework for designing robust learning algorithms in modern signal processing. By employing information-theoretic measures such as correntropy and error entropy, ITL-based methods are able to capture higher-order statistical characteristics of signals and therefore provide improved robustness in non-Gaussian and impulsive noise environments. In recent years, adaptive filtering algorithms developed under the ITL framework have received considerable attention due to their strong robustness and excellent performance in complex signal processing scenarios. Motivated by these advantages, this book focuses on the development, theoretical analysis, and applications of adaptive filtering algorithms based on information theoretic learning.

This book is organized as follows. Chapter 1 serves as the introduction, presenting the research background, application significance of information theoretic learning and adaptive filtering, as well as the main topics studied in this book. Chapter 2 investigates the transient performance of adaptive filtering algorithms based on the maximum correntropy criterion (MCC) and analyzes their convergence behavior and performance in non-Gaussian noise environments. Chapter 3 proposes a decorrelation recursive maximum correntropy criterion (DRMCC) algorithm, which improves the convergence speed and estimation accuracy of adaptive filtering by incorporating a decorrelation mechanism. Chapter 4 introduces the combinatoric correntropy criterion, which enhances the performance of traditional correntropy-based learning methods through a combinational correntropy measure. Chapter 5 develops a fast minimum error entropy algorithm with fiducial points to reduce the computational complexity of the minimum error entropy method while maintaining its learning capability. Chapter 6 proposes the asymmetric correntropy criterion to address signal processing problems in environments with asymmetric noise distributions. Chapter 7 studies a hybrid geometric algebra correntropy method, extending correntropy-based learning into the geometric algebra framework for multidimensional signal processing. Chapter 8 presents a kernel generalized maximum correntropy algorithm with variable centers to improve modeling capability under non-ero-mean noise conditions. Chapter 9 investigates kernel adaptive filtering in distributed networks and proposes a diffusion kernel generalized maximum correntropy algorithm to enhance robustness in complex noise environments. Chapter 10 studies a graph-based diffusion kernel maximum correntropy algorithm for processing graph-structured data in scenarios such as wireless sensor networks. Chapter 11 develops a nonlinear graph diffusion maximum mixture correntropy algorithm for modeling nonlinear graph systems in non-Gaussian noise environments.

This book is written on the basis of the author's many years of teaching and research experience. Its purpose is to provide a systematic introduction to the fundamental theories and commonly used methods in this field. In organizing the material, the book attempts to balance theoretical rigor with practical relevance. While maintaining mathematical accuracy, the presentation emphasizes clear explanations of basic concepts and methodological principles, enabling readers to understand the underlying ideas of the algorithms and their conditions of applicability, and to apply them in practical problems.

This book is primarily intended for senior undergraduate students and graduate students, and it may also serve as a reference for researchers and engineers working in related areas. Readers are expected to have a basic background in mathematics, including linear algebra, probability theory, and digital signal processing. The material is organized in a progressive manner so that readers can gradually build a systematic knowledge framework.

In the writing process, special attention is given to the combination of theoretical analysis and conceptual understanding. Important theoretical derivations are presented with necessary intermediate steps, accompanied by explanations that help readers grasp the essential ideas behind the methods. For several representative techniques, discussions are also provided in the context of practical applications, enabling readers to understand their significance in real engineering problems. In addition, issues such as algorithm performance, computational complexity, and application conditions are discussed where appropriate, so as to enhance readers' overall understanding of the techniques.

The completion of this book has benefited from the support and assistance of many colleagues and students. During long-term teaching and research activities, their comments and suggestions have contributed greatly to the improvement of the manuscript. The author would like to express sincere gratitude to them.

Despite careful preparation, shortcomings and errors may still exist in the book. The author would greatly appreciate comments and suggestions from readers so that improvements can be made in future revisions.

Chengdu, China — Haiquan Zhao
Lanzhou, China — Xinyan Hou
Lanzhou, China — Xiaoqiang Long

Acknowledgments We would like to thank Dr. Xinyan Hou, and Dr. Xiaoqiang Long with whom we have worked on the topic of this book and who contributed to some of the results reported here. This work was partially supported by National Natural Science Foundation of China (grant: 62671522, 62171388, 61871461) and partially supported by 2026 Joint Innovation Fund for Pairing-Assisted Universities (grant: 1200150767).

Competing Interests The authors have no competing interests to declare that are relevant to the content of this manuscript.

Contents

Chapter 1
Introduction

With the rapid advancement of information technology, the efficient processing of massive data has become a critical challenge in defense and engineering technology. Against this backdrop, classical adaptive filtering technology, known for its high processing efficiency, has demonstrated significant application value. Currently, it is widely employed in scenarios such as underwater sonar positioning, radar target direction finding and tracking, and military satellite communication anti-jamming. These applications substantially enhance the target detection and tracking accuracy of military systems in high-interference environments, providing crucial technical support for key defense systems. In civilian applications, adaptive filters are used in active noise-canceling headphones and in-vehicle systems for real-time noise suppression, effectively improving speech communication clarity and audio fidelity. From interference suppression to signal enhancement, adaptive filtering algorithms offer core solutions to complex engineering problems in signal processing, continuously driving advancements in both defense and civilian technologies.

An adaptive filter consists of a digital filter with adjustable parameters and structure, along with an adaptive algorithm, as illustrated in Fig. 1.1. The performance of the adaptive filter is primarily determined by the adaptive filtering algorithm. Adaptive filtering algorithms represent a signal processing technique developed from the theoretical foundations of Wiener and Kalman filtering. Unlike Wiener and Kalman filtering, adaptive filtering algorithms dynamically adjust filter coefficients based on the error between the filter output and the desired output. This enables real-time adaptation to unknown or time-varying environments without relying on prior statistical knowledge of the signal and noise. Common adaptive filtering algorithms are generally categorized into two types based on their structure: linear adaptive filtering and nonlinear adaptive filtering algorithms.

In practical applications, systems are often affected by non-Gaussian noise (such as impulsive noise), leading to significant performance degradation of classical adaptive filtering algorithms designed based on the Minimum Mean Square Error (MMSE)

H. Zhao et al., *Information Theoretic Learning-Based Filter*,
https://doi.org/10.1007/978-3-032-29623-8_1

Fig. 1.1 Adaptive filter architecture block diagram

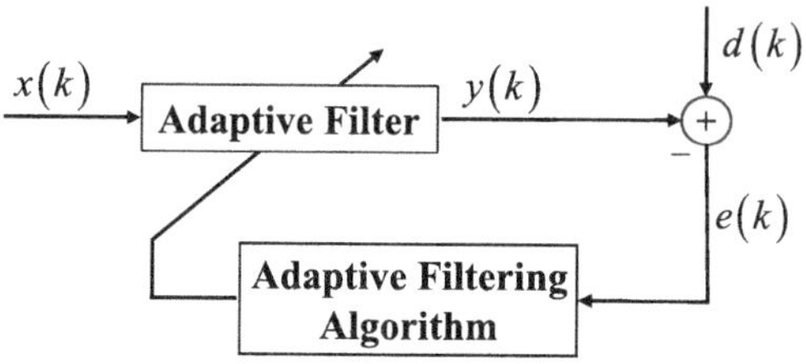

criterion, such as the Least Mean Square (LMS), Affine Projection (AP), and Recursive Least Square (RLS) algorithms. To address this challenge, Príncipe et al. introduced the Information Theoretic Learning (ITL) framework into the field of adaptive filtering, with its core theoretical foundation built upon Rényi entropy. In information theory, Rényi entropy quantifies the uncertainty or randomness of a random variable, revealing the degree of disorder within a system. By integrating entropy measures from information theory to describe data uncertainty, ITL enables the design of highly adaptive learning algorithms. As an important tool in machine learning, Rényi entropy has not only been applied in areas such as blind source separation and feature extraction but has also facilitated the development of core concepts in ITL, including Error Entropy and Correntropy. Since information-theoretic metrics (ITMs) based on Rényi entropy can capture richer statistical information of data, they exhibit strong robustness when dealing with non-Gaussian and impulsive noise. Currently, robust criteria based on ITL have formed a comprehensive methodological framework and have been successfully integrated into the design of both linear and nonlinear adaptive filtering algorithms, significantly improving the accuracy of signal processing. Examples include the Maximum Correntropy (MC) and Minimum Error Entropy (MEE) criteria.

Additionally, several critical issues requiring urgent resolution persist in the practical engineering applications of adaptive filtering algorithms: First, during system identification, non-Gaussian noise interference often leads to a significant degradation in the performance of traditional algorithms; second, noise contamination in input signals severely impacts call quality in speech communication scenarios. Furthermore, existing methods frequently face accuracy limitations in applications such as multidimensional signal prediction (e.g., wind speed forecasting, heterogeneous data prediction) and frequency estimation in three-phase power systems. Effectively resolving these issues will directly enhance the performance of relevant systems: improved adaptive filtering algorithms can increase prediction accuracy and enhance speech communication quality.

This situation has prompted researchers to explore more robust solutions, among which correntropy based on Information Theoretic Learning (ITL) has garnered significant attention due to its outstanding interference resistance. In recent years, building upon ITL, algorithms based on the Maximum Correntropy Criterion (MCC) and its derivatives have been extensively studied, and their effectiveness in handling non-Gaussian system noise has been demonstrated. Compared to conventional AFAs, robust AFAs based on correntropy exhibit significant advantages: not only do they achieve faster convergence and lower steady-state error, but they also demonstrate superior predictive capabilities. These performance advantages are particularly pronounced in scenarios involving non-Gaussian noise environments, noisy input/output signals, and complex system predictions. Therefore, in-depth research on robust AFAs based on correntropy using improved kernel functions not only provides a new theoretical framework and technical pathway to address the performance limitations of traditional algorithms in complex environments but also offers superior solutions for practical engineering applications. This holds significant theoretical and practical value for advancing technological progress in the field of signal processing.

From a theoretical perspective, robust AFAs based on correntropy demonstrate multifaceted performance advantages and hold significant theoretical value: First, in non-Gaussian interference environments, the improved kernel function enables correntropy to capture signal characteristics through higher-order statistics. Its robustness significantly outperforms traditional MMSE algorithms based on second-order statistics, an advantage confirmed through rigorous mathematical derivations and extensive experimental validation. Second, addressing the challenge of noisy input signals, the innovative integration of the improved kernel function's correntropy with a bias compensation mechanism systematically corrects estimation errors caused by input noise, substantially enhancing the algorithm's convergence accuracy and stability. Furthermore, by incorporating multidimensional correntropies (such as geometric algebraic correntropy), the improved kernel function's correlation entropy gains the capability to process multidimensional signals, offering novel solutions for state estimation in complex systems. Notably, in asymmetric noise environments where errors follow skewed distributions, traditional algorithms based on symmetric kernels face performance limitations. The introduction of the maximum asymmetric correntropy criterion based on the improved kernel not only refines research on asymmetric correntropy algorithms but also provides a highly effective tool for such specialized scenarios, demonstrating significant innovative value.

At the engineering application level, robust adaptive filtering algorithms based on correntropy demonstrate unique value and practical significance. These algorithms exhibit outstanding performance in fields such as wind speed prediction, power system frequency estimation, and AEC. Their performance advantages are primarily manifested in the following aspects: First, by improving the kernel function, the proposed algorithm enhances its processing capability for non-stationary signals, achieving more precise real-time estimation in applications like power system frequency estimation and wind speed prediction. Second, the enhanced kernel structure improves the algorithm's adaptability to complex disturbances, ensuring stable performance even in noisy environments. These advancements contribute to the development of correntropy theory in nonlinear signal processing. From a theoretical innovation perspective, the design of the improved kernel function overcomes the limitations of traditional single-kernel correntropy criteria, establishing a more universal signal processing framework. In terms of engineering practicality, the proposed method provides a reliable technical solution for tackling real-time signal processing challenges such as wind speed prediction and power system frequency estimation, holding significant theoretical value and engineering practical significance.

Although ITL-based adaptive filtering algorithms perform well in non-Gaussian noise environments, a complete theoretical framework to accurately predict their transient behavior during iteration has yet to be established. This uncertainty may affect the convergence speed and final performance of the algorithms, thereby impacting their effectiveness in practical applications. Therefore, studying the transient performance analysis of ITL-based adaptive filtering algorithms in non-Gaussian noise environments is crucial for ensuring their effectiveness and practicality. Secondly, adaptive filtering algorithms based on traditional MEE and MC criteria may not always deliver optimal performance in certain environments. Thus, designing corresponding ITL-based criteria tailored to different noise environments is of great significance for improving algorithmic performance. Furthermore, in many practical applications, signals often exhibit nonlinear characteristics, and traditional linear adaptive filtering algorithms may face performance bottlenecks when dealing with such problems. This is because linear models struggle to describe the properties of nonlinear signals, especially in environments with nonlinear distortions and non-Gaussian noise interference. Therefore, researching nonlinear adaptive filtering algorithms to compensate for the performance limitations of linear algorithms in handling such complex signals has become an urgent and critical research topic.

It is worth noting that signals in many real-world scenarios are recorded in interactive structures, and data often exhibit certain correlations, as seen in social networks, transportation networks, and wireless sensor networks. The interconnected nature of such data results in irregular and complex structures. Traditional adaptive signal processing methods are often inadequate for handling such complexities, while Graph Signal Processing (GSP) offers a new approach for processing such data. A graph refers to a mathematical structure composed of a set of vertices and a set of edges that describe the relationships between vertices. Typical graph topologies include Erdős–Rényi graphs, ring graphs, random geometric graphs, and others. Unlike classical signal processing, GSP incorporates graph structures into the signal processing framework, leveraging vertex and edge relationships to characterize complex data correlations, thereby extending classical methods to irregular data domains. Currently, GSP has been widely applied in areas such as sensor networks, image processing, and speech enhancement.

However, several challenges remain. For instance, how to apply concepts from information-theoretic learning to graph signal processing and develop robust nonlinear processing methods based on graph signals remains an unresolved challenge. Secondly, for nonlinear graph signal processing, effective theoretical analysis methods to predict the transient and steady-state behaviors of algorithms are still lacking. Therefore, research on robust nonlinear adaptive filtering algorithms based on graph signals holds significant practical value for addressing these issues. Finally, existing nonlinear adaptive filtering algorithms for graph signals employ various nonlinear extension forms. Establishing a unified framework for these algorithms is of great importance for advancing the field of nonlinear graph signal processing.

Chapter 2
Transient Analysis of Adaptive Filtering Algorithms Based on the Maximum Correntropy Criterion

The adaptive filtering algorithm based on the maximum correntropy criterion (MCC) is very effective in suppressing non-Gaussian noises and therefore attracts widespread attention. At present, some work has been done to study the convergence and steady-state performance analysis of the MCC algorithm, but its transient performance analysis is still an open problem. To provide a comprehensive theoretical foundation for the MCC algorithm, we propose a method for transient performance analysis based on moment generating function (MGF). Since this method can efficiently calculate the expected value of the exponential term in the iterative update equation, it can avoid the discrepancies caused by introducing some approximation methods such as Taylor expansions in the analysis process. To date, there is no precedent for using this method to analyze the transient performance of the MCC algorithm. In addition, the steady-state performance and stability conditions of the MCC algorithm are discussed based on this method.

2.1 Introduction

Many signal processing applications such as hearing aids, noise cancellation, and channel equalization require adaptive filtering. It is well known that a vital issue in adaptive filtering theory and applications is how to design a suitable cost function to meet different application scenarios. The minimum mean square error (MMSE) criterion has been widely used in adaptive filtering due to its convexity, smoothness, simplicity, and optimality under the Gaussian assumption [1, 2]. However, the performance of the algorithm based on the MMSE criterion will be drastically degraded when the data is disturbed by non-Gaussian noise.

Recently, there are various non-MMSE criteria proposed for the improvement of algorithm performance in non-Gaussian environments. The least mean p-power (LMP) algorithm is a typical example, which employs the pth order absolute moment

H. Zhao et al., *Information Theoretic Learning-Based Filter*,
https://doi.org/10.1007/978-3-032-29623-8_2

of the error as a cost function [3, 4]. Based on the high-order error statistics information, the LMP algorithm can improve performance significantly when the desired signal contains light-tailed non-Gaussian noise. However, the low-order error statistics measurement is usually more robust when the desired signal is contaminated by the heavy-tailed impulsive noise [5]. For example, the maximum correntropy criterion (MCC) can effectively attenuate the adverse effects of impulsive or non-Gaussian noises [6–15]. Since the MCC algorithm may not achieve the desired results in some specific scenarios, scholars have improved the MCC algorithm from different perspectives. In order to improve the flexibility, the maximum mixture correntropy criterion (MMCC) is proposed, which combines two Gaussian functions with distinct kernel widths as cost functions and achieves better performance than the conventional MCC algorithm [16]. In addition, the standard zero-mean Gaussian function usually does not match the error distribution well when the mean value of the error distribution is not zero. To overcome this dilemma, the MCC with variable center (MCC-VC) algorithm is proposed [17]. To address the problem that the algorithm performance deteriorates when the input signal contains noise, a gradient-based maximum total correntropy (MTC) algorithm is proposed, which can well estimate the error-in-variables model [18].

Although the MCC algorithm and its variants may achieve outstanding performance in non-Gaussian noise, at present, most scholars focus on the steady-state performance analysis of the MCC algorithm [8, 19], and its transient performance analysis is still pending. Some existing analytical methods of MCC-type algorithms mainly focus on fixed-point adaptive filtering algorithms. For example, a detailed study of the transient behavior of the fixed-point MTC algorithm has been reported based on the statistical average approximation method [20]. Unfortunately, this method requires the expected value of the exponential term in the iterative update equation to be predetermined, and it is not directly applicable to the transient behavior analysis of the MCC algorithm. Inspired by the moment generating function (MGF), an innovative transient analysis method for MCC is proposed in this chapter, and the closed-form expression of mean square deviation (MSD) is also provided. Finally, the simulation results verify the accuracy of the proposed theoretical model.

2.2 Maximum Correntropy Criterion (MCC)

Correntropy, a nonlinear measure of the similarity between two random variables $\mathcal{A}$ and $\mathcal{B}$, is defined as

$$V(\mathcal{A}, \mathcal{B}) = E[k(\mathcal{A}, \mathcal{B})] \tag{2.1}$$

where $k(\cdot, \cdot)$ is a shift-invariant Mercer kernel function, the Gaussian kernel is the most commonly used Mercer kernel function,

$$k(\mathcal{A}, \mathcal{B}) = \exp\left(-\frac{(\mathcal{A} - \mathcal{B})^2}{2\sigma^2}\right) \tag{2.2}$$

where σ is kernel bandwidth. Combining with the Gaussian kernel function, the cost function of the MCC algorithm is defined as

$$J_{\text{MCC}} = E\left[\exp\left(-\frac{e^2(m)}{2\sigma^2}\right)\right] \tag{2.3}$$

where $e(m)$ is the prediction error,

$$e(m) = d(m) - \boldsymbol{w}^{\text{T}}(m)\boldsymbol{x}(m) \tag{2.4}$$

where $(\cdot)^{\text{T}}$ denotes the transpose operator, $d(m) \in \mathbb{R}$ is the desired signal, $\boldsymbol{x}(m) \in \mathbb{R}^L$ and $\boldsymbol{w}(m) \in \mathbb{R}^L$ are the input vector and weight vector of the adaptive filter at time m, respectively.

Suppose the desired value $d(m)$ is denoted as

$$d(m) = \boldsymbol{w}_{\text{opt}}^{\text{T}}\boldsymbol{x}(m) + v(m) \tag{2.5}$$

where $\boldsymbol{w}_{\text{opt}} \in \mathbb{R}^L$ is the system parameter, and $v(m)$ is the measurement noise.

2.3 Adaptive Filtering Based on Maximum Correntropy Criterion

The optimal weight vector of the MCC algorithm is obtained by maximizing the cost function J_{MCC}. Combined with the stochastic gradient method, the weight update equation of the MCC algorithm is represented as

$$\boldsymbol{w}(m+1) = \boldsymbol{w}(m) + \mu e(m)\exp\left(-\frac{e^2(m)}{2\sigma^2}\right)\boldsymbol{x}(m) \tag{2.6}$$

where μ is the step size of the MCC algorithm. One can see that the MCC algorithm contains an exponential factor in the weight update Eq. (2.6), which can effectively attenuate the adverse effects of impulsive noise and outliers, compared with the least mean square (LMS) algorithm that contains only second-order statistical information.

2.4 Theoretical Analysis

To facilitate the performance analysis of the MCC algorithm, the below assumptions are employed:

A1 The measurement noise $v(m)$ is independent, identically distributed (i.i.d.) process with zero mean and variance $\sigma_v^2 = E\left[v^2(m)\right]$.
A2 The input vector $\boldsymbol{x}(m)$ is zero-mean independent Gaussian process and its covariance matrix $\boldsymbol{R}_{xx} = E\left[\boldsymbol{x}(m)\boldsymbol{x}^{\mathrm{T}}(m)\right]$ is positive definite.
A3 The measurement noise $v(m)$, the input vector $\boldsymbol{x}(m)$ and the MGF of $e^2(m)$ are mutually independent.

2.4.1 Mean Weight Behavior

Defining $\hat{\boldsymbol{w}}(m) = \boldsymbol{w}(m) - \boldsymbol{w}_{\mathrm{opt}}$ as the weight-error vector, the estimation error $e(m)$ can be written as

$$e(m) = v(m) - \hat{\boldsymbol{w}}^{\mathrm{T}}(m)\boldsymbol{x}(m) \tag{2.7}$$

Combining (2.6) and (2.7), the update equation of weight-error vector $\hat{\boldsymbol{w}}(m+1)$ is obtained,

$$\begin{aligned}\hat{\boldsymbol{w}}(m+1) = {} & \mu \exp\left(-\frac{e^2(m)}{2\sigma^2}\right) v(m)\boldsymbol{x}(m) \\ & + \left[\boldsymbol{I}_L - \mu \exp\left(-\frac{e^2(m)}{2\sigma^2}\right)\boldsymbol{x}(m)\boldsymbol{x}^{\mathrm{T}}(m)\right]\hat{\boldsymbol{w}}(m)\end{aligned} \tag{2.8}$$

where $\boldsymbol{I}_L$ is the $L \times L$ identity matrix.

Remark 2.1 To simplify the analysis process, we did not discuss the case of correlated input sequences. In fact, the correlated input sequence problem can be solved by the method of eigenvalue decomposition [21], i.e.,

$$\boldsymbol{\Lambda} = \boldsymbol{A}^{\mathrm{T}}\boldsymbol{R}_{xx}\boldsymbol{A}$$

where $\boldsymbol{A} \in \mathbb{R}^{L\times L}$ is the orthogonal matrix and $\boldsymbol{\Lambda} \in \mathbb{R}^{L\times L}$ is the diagonal matrix composed of eigenvalues of $\boldsymbol{R}_{xx}$. The subsequent analysis is similar to the uncorrelated input sequences issue discussed in this chapter.

Taking expectations of (2.8) and using assumptions A1–A3 yields

$$E[\hat{\boldsymbol{w}}(m+1)] = \left[I_L - \mu E\left[\exp\left(-\frac{e^2(m)}{2\sigma^2}\right)\right]\boldsymbol{R}_{xx}\right]E[\hat{\boldsymbol{w}}(m)] \tag{2.9}$$

For subsequent derivation, the following notations are introduced.

$$\boldsymbol{y}(m) = \left[\, v(m)\ \hat{\boldsymbol{w}}^{\mathrm{T}}(m)\boldsymbol{x}(m)\,\right]^{\mathrm{T}} \tag{2.10}$$

$$\boldsymbol{Q} = \begin{bmatrix} 1 & -1 \\ -1 & 1 \end{bmatrix} \tag{2.11}$$

where $\boldsymbol{y}(m)$ is a zero mean vector according to assumptions A1 and A2. Therefore, the square of the prediction error can be written in quadratic form.

$$e^2(m) = \boldsymbol{y}^{\mathrm{T}}(m)\boldsymbol{Q}\boldsymbol{y}(m) \tag{2.12}$$

Employing assumptions A1–A3, we obtain the covariance matrix of $\boldsymbol{y}(m)$, i.e.,

$$\boldsymbol{R}_y(m) = E\left[\boldsymbol{y}(m)\boldsymbol{y}^{\mathrm{T}}(m)\right] = \begin{bmatrix} \sigma_v^2 & 0 \\ 0 & \mathrm{Tr}(\boldsymbol{R}_{xx}\boldsymbol{W}(m)) \end{bmatrix} \tag{2.13}$$

where $\mathrm{Tr}(\cdot)$ is trace of the matrix, $\mathrm{Tr}(\boldsymbol{R}_{xx}\boldsymbol{W}(m))$ is excess mean square error (EMSE), and $\boldsymbol{W}(m) = E\left[\hat{\boldsymbol{w}}(m)\hat{\boldsymbol{w}}^{\mathrm{T}}(m)\right]$ is the autocorrelation matrix of $\hat{\boldsymbol{w}}(m)$.

Regarding the measurement noise $v(m)$ obeying different distributions, the calculation of the MGF of quadratic form $z(m) = e^2(m)$ is as follows.

(1) When the measurement noise $v(m)$ obeys Gaussian distribution, the MGF of $z(m)$ can be calculated as [22, pp. 100]

$$\psi_{z(m)}(s) = E\left[\exp(sz(m))\right] = \det\left(\boldsymbol{I}_2 - 2s\boldsymbol{Q}\boldsymbol{R}_y(m)\right)^{-\frac{1}{2}} \tag{2.14}$$

where $\det(\cdot)$ denotes the determinant of the matrix and s is the variable. By setting $s = -\frac{1}{2\sigma^2}$ in (2.14), the mathematical expectation of $\exp\left(-\frac{e^2(m)}{2\sigma^2}\right)$ can be given by

$$E\left[\exp\left(-\frac{e^2(m)}{2\sigma^2}\right)\right] = \psi_{z(m)}\left(-\frac{1}{2\sigma^2}\right) = \det\left(\boldsymbol{I}_2 + \boldsymbol{Q}\boldsymbol{R}_y(m)/\sigma^2\right)^{-\frac{1}{2}} \tag{2.15}$$

When the measurement noise $v(m)$ is non-Gaussian, $\psi_{z(m)}$ can be approximated as [23]

$$\psi_{z(m)}(s) \approx \psi_{V(m)}(s)\psi_{X(m)}(s) \tag{2.16}$$

where $X(m) = \hat{\boldsymbol{w}}^{\mathrm{T}}(m)\boldsymbol{x}(m)\boldsymbol{x}^{\mathrm{T}}(m)\hat{\boldsymbol{w}}(m)$, $\psi_{X(m)}(s) = (1 - 2s\mathrm{Tr}(\boldsymbol{R}_{xx}\boldsymbol{W}(m)))^{-\frac{1}{2}}$ and $V(m) = v^2(m)$. The MGFs of the square $V(m)$ of the measurement noise obeying Gaussian, Uniform and Binary distributions are summarized in Table 2.1 [23], where $\mathrm{erf}\left(a\sqrt{-s}\right) = \frac{2}{\sqrt{\pi}}\int_0^{a\sqrt{-s}} \exp\left(-t^2\right)\mathrm{d}t$ is the error function.

According to the above conclusion, (2.9) can be rewritten as

Table 2.1 MGF of the square of measurement noise

Distributions	MGF
$v(m) \sim N(0, \rho^2)$	$\psi_{V(m)}(s) = \sqrt{\frac{1}{1-2s\rho^2}}$
$v(m) \sim B(1, p)$	$\psi_{V(m)}(s) = pe^s + 1 - p$
$v(m) \sim U(-a, a)$	$\psi_{V(m)}(s) = \frac{\sqrt{\pi}\,\mathrm{erf}(a\sqrt{-s})}{2a\sqrt{-s}}$

$$E[\hat{\boldsymbol{w}}(m+1)] = [I_L - \eta_m \boldsymbol{R}_{xx}]E[\hat{\boldsymbol{w}}(m)] \tag{2.17}$$

where

$$\eta_m = \mu \psi_{z(m)}\left(-\frac{1}{2\sigma^2}\right) \tag{2.18}$$

Thus, the MCC algorithm can be regarded as a variable step size LMS algorithm. When $\sigma \to \infty$ in (2.15), the MCC algorithm will degrade to the LMS algorithm in Gaussian noise. If η_m is obtained, the transient and steady-state behaviors of the MCC algorithm can be easily predicted. Also important is that, distinct from the literature [20], we do not assume that η_m is a constant, and this method is also applicable to the analysis of other types of MCC algorithms, such as MMCC, MCC-VC, and MTC.

To obtain the transient behavior of each weight in mean sense, (2.17) is rewritten as

$$E[\boldsymbol{w}(m+1)] = [I_L - \eta_m \boldsymbol{R}_{xx}]E[\boldsymbol{w}(m)] + \eta_m \boldsymbol{R}_{xx}\boldsymbol{w}_{opt} \tag{2.19}$$

According to (2.17), for any initial condition, the MCC algorithm (2.6) converges asymptotically if the step size satisfies

$$0 < \mu < \frac{2}{\lambda_{\max}\left(\psi_{z(m)}\left(-\frac{1}{2\sigma^2}\right)\boldsymbol{R}_{xx}\right)} \tag{2.20}$$

where $\lambda_{\max}(\cdot)$ denotes the maximum eigenvalue of the matrix.

2.4.2 Second-Order Moment Behavior

Mean weight analysis is not always practical because adaptive filters that are stable in the mean sense may diverge in practice due to deviations from the unbounded variance of the adaptive weight vector. Alternatively, second-order moment analysis can be used to evaluate the MSD defined as

$$\mathrm{MSD}(m) = E\left[\|\hat{\boldsymbol{w}}(m)\|^2\right] = \mathrm{Tr}(\boldsymbol{W}(m)) \tag{2.21}$$

where $\|\cdot\|^2$ denotes the Euclidean norm square. Post-multiplying (8) by its transpose, using assumptions A1–A3 and taking expectations, it leads to

$$\begin{aligned}\boldsymbol{W}(m+1) = \boldsymbol{W}(m) + \eta_m^2\sigma_v^2\boldsymbol{R}_{xx} + \eta_m^2 E\big[\boldsymbol{x}(m)\boldsymbol{x}^{\mathrm{T}}(m)\hat{\boldsymbol{w}}(m)\hat{\boldsymbol{w}}^{\mathrm{T}}(m)\boldsymbol{x}(m)\boldsymbol{x}^{\mathrm{T}}(m)\big] \\ - \eta_m[\boldsymbol{W}(m)\boldsymbol{R}_{xx} + \boldsymbol{R}_{xx}\boldsymbol{W}(m)]\end{aligned} \tag{2.22}$$

Using the Gaussian moment factorization theorem (also known as Isserlis' theorem) [24], the fourth-order moments in (2.22) can be decomposed as

$$E\big[\boldsymbol{x}(m)\boldsymbol{x}^{\mathrm{T}}(m)\hat{\boldsymbol{w}}(m)\hat{\boldsymbol{w}}^{\mathrm{T}}(m)\boldsymbol{x}(m)\boldsymbol{x}^{\mathrm{T}}(m)\big] = 2\boldsymbol{R}_{xx}\boldsymbol{W}(m)\boldsymbol{R}_{xx} + \boldsymbol{R}_{xx}\mathrm{Tr}(\boldsymbol{R}_{xx}\boldsymbol{W}(m)) \tag{2.23}$$

To obtain the transient behavior of $\boldsymbol{W}(m)$ in mean square sense, the following Kronecker product operation and its property are considered in this chapter.

$$\mathrm{vec}(\boldsymbol{UVW}) = \big[\boldsymbol{W}^{\mathrm{T}} \otimes \boldsymbol{U}\big]\mathrm{vec}(\boldsymbol{V}) \tag{2.24}$$

$$\mathrm{Tr}(\boldsymbol{UV}) = \big(\mathrm{vec}\big(\boldsymbol{U}^{\mathrm{T}}\big)\big)^{\mathrm{T}}\mathrm{vec}(\boldsymbol{V}) \tag{2.25}$$

Substituting (2.23), (2.24), and (2.25) into (2.22) yields

$$\begin{aligned}\mathrm{vec}(\boldsymbol{W}(m+1)) = \mathrm{vec}(\boldsymbol{W}(m)) - \eta_m[\boldsymbol{I} \otimes \boldsymbol{R}_{xx}]\mathrm{vec}(\boldsymbol{W}(m)) \\ - \eta_m\big[\boldsymbol{R}_{xx}^{T} \otimes \boldsymbol{I}\big]\mathrm{vec}(\boldsymbol{W}(m)) + 2\eta_m^2\big[\boldsymbol{R}_{xx}^{\mathrm{T}} \otimes \boldsymbol{R}_{xx}\big]\mathrm{vec}(\boldsymbol{W}(m)) \\ + \eta_m^2\mathrm{vec}(\boldsymbol{R}_{xx})\mathrm{vec}\big(\boldsymbol{R}_{xx}^{\mathrm{T}}\big)^{\mathrm{T}}\mathrm{vec}(\boldsymbol{W}(m)) + \eta_m^2\sigma_v^2\mathrm{vec}(\boldsymbol{R}_{xx})\end{aligned} \tag{2.26}$$

Then, the MSD at time m can be computed by $\mathrm{vec}(\boldsymbol{I}_L)^{\mathrm{T}}\mathrm{vec}(\boldsymbol{W}(m))$.

2.4.3 Mean Square Steady-State Behavior

The steady-state performance of the MCC algorithm is analyzed in this subsection. The filter is hypothesized to be stable and reach a steady state after a while, such that the autocorrelation matrix of the weight-error vector satisfies

$$\lim_{m\to\infty}\boldsymbol{W}(m) = \lim_{m\to\infty}\boldsymbol{W}(m+1) \tag{2.27}$$

Then the steady-state equation from (2.26) can be written as

$$\begin{aligned}\eta_\infty\big[(\boldsymbol{I} \otimes \boldsymbol{R}_{xx})\mathrm{vec}(\boldsymbol{W}(\infty)) + \big(\boldsymbol{R}_{xx}^{\mathrm{T}} \otimes \boldsymbol{I}\big)\mathrm{vec}(\boldsymbol{W}(\infty))\big] \\ - 2\eta_\infty^2\big[\boldsymbol{R}_{xx}^{\mathrm{T}} \otimes \boldsymbol{R}_{xx}\big]\mathrm{vec}(\boldsymbol{W}(\infty)) - \eta_\infty^2\mathrm{vec}(\boldsymbol{R}_{xx})\mathrm{vec}\big(\boldsymbol{R}_{xx}^{T}\big)^{\mathrm{T}}\mathrm{vec}(\boldsymbol{W}(\infty)) \\ = \eta_\infty^2\sigma_v^2\mathrm{vec}(\boldsymbol{R}_{xx})\end{aligned} \tag{2.28}$$

In the steady state, we can assume that $\mathrm{Tr}(\boldsymbol{R}_{xx}\boldsymbol{W}(\infty)) \ll \sigma_v^2$ [21] and $\mathrm{Tr}(\boldsymbol{R}_{xx}\boldsymbol{W}(\infty)) \approx 0$ holds, and substituting it into (2.16) leads to $\psi_{X(\infty)} = 1$. Thus, η_∞ is obtained as

$$\eta_\infty = \mu\psi_{V(\infty)} \tag{2.29}$$

By solving (2.28), we have

$$\mathrm{vec}(\boldsymbol{W}(\infty)) = \boldsymbol{P}^{-1}\eta_\infty^2\sigma_v^2\mathrm{vec}(\boldsymbol{R}_{xx}) \tag{2.30}$$

where

$$\boldsymbol{P} = \eta_\infty\left[(\boldsymbol{I}\otimes\boldsymbol{R}_{xx}) + \left(\boldsymbol{R}_{xx}^T\otimes\boldsymbol{I}\right)\right] - \eta_\infty^2\left[2\left(\boldsymbol{R}_{xx}^T\otimes\boldsymbol{R}_{xx}\right) + \mathrm{vec}(\boldsymbol{R}_{xx})\mathrm{vec}\left(\boldsymbol{R}_{xx}^T\right)^{\mathrm{T}}\right] \tag{2.31}$$

2.5 Computer Simulation Examples

In this section, the system identification problem is used to verify the effectiveness and accuracy of the proposed analysis method, where the desired value $d(m)$ is generated by $d(m) = \boldsymbol{w}_{\mathrm{opt}}^{\mathrm{T}}\boldsymbol{x}(m) + v(m)$. In the simulations, the input signal $\boldsymbol{x}(m)$ is a zero-mean Gaussian white process with the covariance matrix of the identity matrix, the parameter $\boldsymbol{w}_{\mathrm{opt}}$ is created randomly, and the initial weight vector of the adaptive filter is set as null. In addition, the simulated learning curves of the algorithm are obtained by averaging 50 independent Monte Carlo simulations.

First, to evaluate the evolution of the mean behaviors of all weight values of MCC algorithm, the case where the filter length $L = 5$ and the observation noise obeys Gaussian distribution with zero mean and variance 0.1 is considered. The kernel bandwidth σ is 4 and the step size μ is 0.03. One can see from Fig. 2.1 that the theoretical learning curves generated by (19) agree with the Monte Carlo simulation.

Second, we discuss the transient and steady-state behaviors of MSD when the measurement noise obeys different types of distributions. In the simulation, the filter length L is 20 and kernel bandwidth σ is 4. In addition, three noises with different distributions are considered: (1) zero-mean Gaussian distribution with variance σ_v^2; (2) Binary distribution over $\{-p, 1-p\}$ with the probability mass function with $Pr(v = 1-p) = p$. This setting is used to generate zero-mean noise; (3) Uniform distribution over $\{-a, a\}$. Figures 2.2, 2.3 and 2.4 show the transient and steady-state behaviors of the MSD with different step sizes and noise variances when the noise obeys different distributions, respectively. In the figures, the densely dashed line indicates the Monte Carlo simulation, and the solid and dashed lines indicate the transient and steady-state behaviors obtained from the corresponding theoretical models (2.26) and (2.30). As shown in Figs. 2.2, 2.3 and 2.4, the theoretical learning

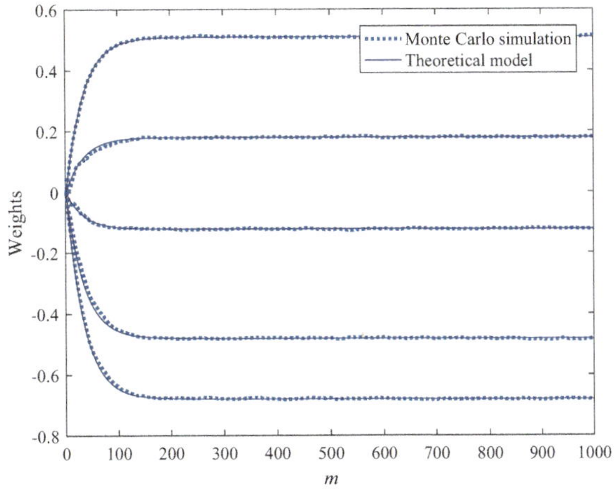

Fig. 2.1 Theoretical and simulated learning curves of all weight values for MCC algorithm

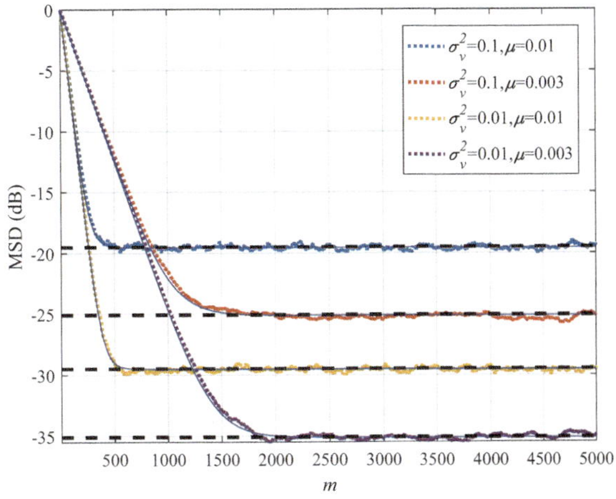

Fig. 2.2 Transient and steady-state behaviors for the MCC algorithm over various step size and variance in Gaussian noise

curves of MSD agree well with their corresponding Monte Carlo simulation curves in both transient and steady-state.

2.6 Summary

In this chapter, we focus on the transient and steady-state behaviors of the MCC algorithm in the mean and mean-square sense when the measurement noise obeys different distributions. The excellent agreement between the theoretical model and Monte Carlo simulations verifies the correctness of the proposed model. It is notable that the method can be used to predict the behavior of the MCC algorithm when the

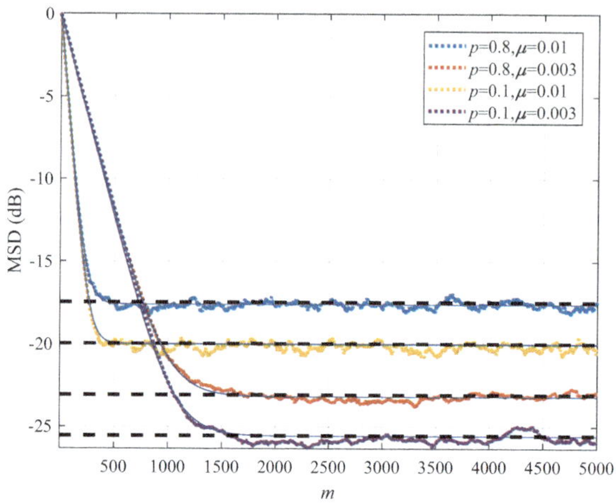

Fig. 2.3 Transient and steady-state behaviors for the MCC algorithm over various step size and variance in Binary noise

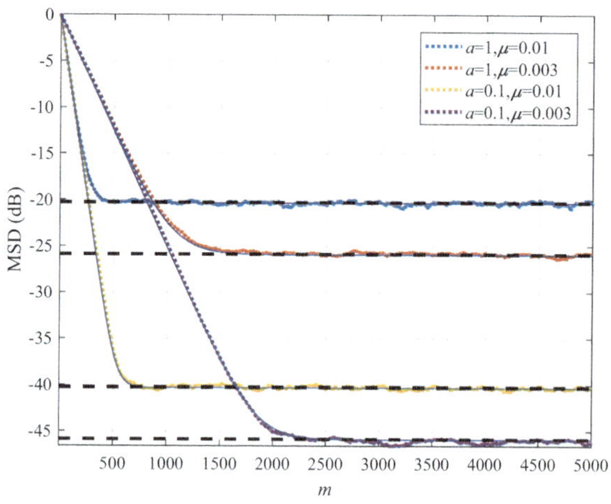

Fig. 2.4 Transient and steady-state behaviors for the MCC algorithm over various step size and variance in uniform noise

MGF of the $v^2(m)$ exists. In future work, we will further explore the performance analysis of other MCC-type algorithms.

References

1. S. Haykin, *Adaptive Filter Theory*, 3rd edn. (Prentice-Hall, Englewood Cliffs, NJ, USA, 1996)
2. A.H. Sayed, *Fundamentals of Adaptive Filtering* (Wiley, Hoboken, NJ, USA, 2003)
3. S. Pei, C. Tseng, Least mean p-power error criterion for adaptive fir filter. IEEE J. Sel. Areas Commun. **12**(9), 1540–1547 (1994)
4. Y. Xiao, Y. Tadokoro, K. Shida, Adaptive algorithm based on least mean p-power error criterion for Fourier analysis in additive noise. IEEE Trans. Signal Process. **47**(4), 1172–1181 (1999)

5. S. P. Talebi, S. Werner, S. Li, D. P. Mandic, Tracking dynamic systems in α-stable environments, in *ICASSP 2019—2019 IEEE International Conference on Acoustics, Speech and Signal Processing (ICASSP)* (2019), pp. 4853–4857
6. W. Liu, P.P. Pokharel, J.C. Principe, Correntropy: properties and applications in non-Gaussian signal processing. IEEE Trans. Signal Process. **55**(11), 5286–5298 (2007)
7. L. Bako, Robustness analysis of a maximum correntropy framework for linear regression. Automatica **87**, 218–225 (2018)
8. B. Lin, R. He, L. Song, B. Wang, Steady-state performance analysis for adaptive filters with error nonlinearities, in *2009 IEEE International Conference on Acoustics, Speech and Signal Processing* (2009), pp. 3093–3096
9. A. Singh, J. C. Principe, Using correntropy as a cost function in linear adaptive filters, in *2009 International Joint Conference on Neural Networks* (2009), pp. 2950–2955
10. L. Shi, Y. Lin, Convex combination of adaptive filters under the maximum correntropy criterion in impulsive interference. IEEE Signal Process. Lett. **21**(11), 1385–1388 (2014)
11. B. Chen, L. Xing, J. Liang, N. Zheng, J.C. Principe, Steady-state mean-square error analysis for adaptive filtering under the maximum correntropy criterion. IEEE Signal Process. Lett. **21**(7), 880–884 (2014)
12. B. Chen, J. Wang, H. Zhao, N. Zheng, J.C. Principe, Convergence of a fixed-point algorithm under maximum correntropy criterion. IEEE Signal Process. Lett. **22**(10), 1723–1727 (2015)
13. H. Zhao, B. Liu, P. Song, Variable step-size affine projection maximum correntropy criterion adaptive filter with correntropy induced metric for sparse system identification. IEEE Trans. Circuits Syst. II Express Briefs **67**(11), 2782–2786 (2020)
14. L. Shi, H. Zhao, Y. Zakharov, An improved variable kernel width for maximum correntropy criterion algorithm. IEEE Trans. Circuits Syst. II Express Briefs **67**(7), 1339–1343 (2020)
15. Y. Zhu, H. Zhao, X. Zeng, B. Chen, Robust generalized maximum correntropy criterion algorithms for active noise control. IEEE/ACM Trans. Audio Speech Lang. Process. **28**, 1282–1292 (2020)
16. B. Chen, X. Wang, N. Lu, S. Wang, J. Cao, J. Qin, Mixture correntropy for robust learning. Pattern Recogn. **79**, 318–327 (2018)
17. B. Chen, X. Wang, Y. Li, J.C. Principe, Maximum correntropy criterion with variable center. IEEE Signal Process. Lett. **26**(8), 1212–1216 (2019)
18. F. Wang, Y. He, S. Wang, B. Chen, Maximum total correntropy adaptive filtering against heavy-tailed noises. Signal Process. **141**, 84–95 (2017)
19. W. Wang, H. Zhao, K. Doğançay, Y. Yu, L. Lu, Z. Zheng, Robust adaptive filtering algorithm based on maximum correntropy criteria for censored regression. Signal Process. **160**, 88–98 (2019)
20. G. Qian, J. Mei, H.H.C. Iu, S. Wang, Fixed-point maximum total complex correntropy algorithm for adaptive filter. IEEE Trans. Signal Process. **69**, 2188–2202 (2021)
21. R. Kwong, E. Johnston, A variable step size LMS algorithm. IEEE Trans. Signal Process. **40**(7), 1633–1642 (1992)
22. J. Omura, T. Kailath, Some useful probability distributions, The Stanford Electronics Laboratories, Stanford University, Stanford, CA, Technical Report 7050-6 (1965)
23. D. P. Bertsekas, J. N. Tsitsiklis, *Introduction to Probability* (Athena Scientific, 2008)
24. V. Mathews, Z. Xie, A stochastic gradient adaptive filter with gradient adaptive step size. IEEE Trans. Signal Process. **41**(6), 2075–2087 (1993)

Chapter 3
Decorrelation Recursive Maximum Correntropy Criterion

Recently, the Bayesian decorrelation algorithms have gained attention because they can effectively avoid trade-off between learning rate and estimation accuracy by circumventing the problems associated with constant step-size and significantly enhance the convergence speed of adaptive filtering algorithms by decorrelation of signals. However, the present Bayesian decorrelation algorithms are constructed on the basis of the minimum-mean-square-error (MMSE) criterion, leading to deterioration in performance when confronted with non-Gaussian noise environments. Therefore, this paper introduces the maximum correntropy criterion in the decorrelation algorithm and develops a decorrelation recursive maximum correntropy criterion (DRMCC) algorithm based on the observation model to achieve robust performance. The proposed algorithm has a similar iterative equation as the Bayesian decorrelation algorithm. In addition, we present the convergence condition and analyze the transient behavior of the DRMCC algorithm. When the free parameters of the proposed algorithm are unknown, a common method of parameter estimation is provided. Finally, numerical simulations corroborate that the designed DRMCC algorithm can realize outstanding performance under system identification and acoustic echo cancellation (AEC) applications in non-Gaussian noise environments and the derived theoretical model can arrive at good agreement with simulation results for both stationary and nonstationary scenarios.

3.1 Introduction

Adaptive filtering algorithms (AFAs) have been extensively employed across several fields, including acoustic echo cancellation (AEC), and active noise control [1, 2]. Among them, the least-mean-square (LMS) algorithm and its normalized variant (normalized LMS, NLMS) gain popularity owing to their simple implementation and low complexity. However, when dealing with strongly correlated input signals like

H. Zhao et al., *Information Theoretic Learning-Based Filter*,
https://doi.org/10.1007/978-3-032-29623-8_3

speech data, the LMS and NLMS algorithms may exhibit reduced convergence speed. To accelerate the algorithm's convergence rate, various frameworks are employed for adaptive filters, including the subband adaptive filter (SAF) [3, 4], the affine projection (AP) [5, 6], and the decorrelation filter [7, 8]. The SAF algorithm can decompose the fullband input and output signals into several mutually independent subbands using the analysis filter bank. This decomposition serves to diminish the correlation among the signals. The AP algorithm effectively mitigates the effects of correlated inputs by combining signals from both current and past time instances. The AP algorithm exhibits high computational complexity because it involves the calculation of the inverse of the input matrix. The decorrelation filter based on observation model utilizes different methods to estimate the vector of decorrelation coefficients of the filter, such as decorrelation NLMS (DNLMS) [8]. However, the DNLMS algorithm suffers from the contradiction between the requirements of adaptation speed and filtering accuracy under constant step-size. To solve this problem, a Bayesian decorrelation LMS (BDLMS) algorithm is proposed by utilizing Bayesian techniques to estimate the coefficients of the system model and the autoregressive model [9].

Unfortunately, the algorithms mentioned above are constructed on the minimum-mean-square-error (MMSE) criterion and encounter issues related to convergence and stability problems when confronted with non-Gaussian noise environments. The widespread occurrence of non-Gaussian noise has sparked significant interest in exploring the non-MMSE criteria to enhance the algorithm's robustness [10]. Over the past years, there has been a notable adoption of information-theoretic learning (ITL) to develop various robust AFAs. This is attributed to its ability to excerpt more statistical information from the original signal [11]. The maximum correntropy criterion (MCC) as a representative of the ITL theory exhibits exceptional performance, especially in impulsive noise environments [12, 13].

To strengthen the robustness of the BDLMS algorithm, a decorrelation recursive MCC (DRMCC) algorithm is developed in this paper, which is developed via incorporating the observation model and MCC criterion. The proposed algorithm is the same as the BDLMS algorithm when the kernel widths tend to infinity. Additionally, the DRMCC algorithm's convergence condition and transient behavior are subjected to analysis. Numerical simulations exhibit that compared to the BDLMS algorithm, the developed DRMCC algorithm retains the advantages of the BDLMS algorithm in processing colored input signals and at the same time exhibits excellent robustness under non-Gaussian noise environments. The contributions of this work are summarized as follows.

(1) The DRMCC algorithm is proposed based on the observation model to cope with the performance degradation of existing decorrelation algorithms in non-Gaussian noise environments.
(2) The convergence conditions and transient behavior of the proposed algorithm are analyzed based on statistical properties.

(3) The experimental results show that the proposed algorithm has excellent performance in system identification and acoustic echo cancellation (AEC) applications.

3.2 Decorrelation Recursive Maximum Correntropy Criterion

Consider the scenario in which the input signal $\mathbf{x}_l$ and the desired signal y_l satisfy the linear model

$$d(m) = \mathbf{x}^T(m)\mathbf{w}_o(m) + v(m) \tag{3.1}$$

where $\mathbf{w}_o(m) \in \mathbb{R}^L$ denotes the system weight vector of length L to be estimated, $v(m)$ is the measurement noise with variance σ_v^2, and $\mathbf{x}(m) = [x(m), \ldots, x(m-L+1)]^T$. In this paper, we assume that $\mathbf{w}_o(m)$ is a random walk model, i.e.,

$$\mathbf{w}_o(m) = \mathbf{w}_o(m-1) + \mathbf{q}(m) \tag{3.2}$$

where $\mathbf{q}(m)$ denotes the process perturbation with covariance matrix $\sigma_q^2\mathbf{I}_L$, and $\mathbf{I}_L$ denotes the $L \times L$ identity matrix. We make the assumption that the correlated input signal $x(m)$ is produced based on the following autoregressive (AR) model:

$$x(m) = \sum_{i=1}^{M} \alpha_i x(m-i) + z(m) = \mathbf{x}_\alpha^T(m)\boldsymbol{\alpha} + z(m) \tag{3.3}$$

where $\mathbf{x}_\alpha(m) = [x(m-1), \ldots, x(m-M)]^T$, $\boldsymbol{\alpha} = [\alpha_1, \ldots, \alpha_M]^T$, $z(m)$ represents a zero-mean noise with variance σ_z^2, and M denotes the order of the autoregressive model. Therefore, Eq. (3.3) is rewritten as

$$\mathbf{x}(m) = \sum_{i=1}^{M} \alpha_i \mathbf{x}(m-i) + \mathbf{z}(m) = \mathbf{X}(m)\boldsymbol{\alpha} + \mathbf{z}(m) \tag{3.4}$$

where $\mathbf{X}(m) = [\mathbf{x}(m-1), \ldots, \mathbf{x}(m-M)]$, and $\mathbf{z}(m) = [z(m), \ldots, z(m-L+1)]^T$. To obtain the decorrelated observation model, we represent the estimate of $\boldsymbol{\alpha}$ at moment m as $\hat{\alpha}(m)$. The decorrelated input signal vector $\overline{\mathbf{x}}(m)$ can be defined as the estimation $\hat{\mathbf{z}}(m)$ of the white noise process $\mathbf{z}(m)$, which is represented as

$$\overline{\mathbf{x}}(m) \triangleq \hat{\mathbf{z}}(m) = \mathbf{x}(m) - \mathbf{X}(m)\hat{\alpha}(m) \tag{3.5}$$

Substituting Eq. (3.5) into Eq. (3.1), we have

$$d(m) = \left[\mathbf{x}^T(m-1)\boldsymbol{w}_o(m), \ldots, \mathbf{x}^T(m-M)\boldsymbol{w}_o(m)\right]\hat{\alpha}(m) + \overline{\mathbf{x}}^T(m)\boldsymbol{w}_o(m) + v(m) \tag{3.6}$$

Under stationary scenario, it holds that $\boldsymbol{w}_o = \boldsymbol{w}_o(m)$, which implies

$$\mathbf{x}^T(m-i)\boldsymbol{w}_o(m) = d(m-i) - v(m-i) \tag{3.7}$$

Under the nonstationary scenario, we assume that the unknown weight vector undergoes gradual change and M is relatively small. This permits us to make the following approximation:

$$\mathbf{x}^T(m-i)\boldsymbol{w}_o(m) \approx d(m-i) - v(m-i) \tag{3.8}$$

Using Eqs. (3.7) and (3.8), Eq. (3.6) is concluded as

$$\begin{aligned} d(m) &\approx [d(m-1) - v(m-1), \ldots, d(m-M) - v(m-M)]\hat{\alpha}(m) + \overline{\mathbf{x}}^T(m)\boldsymbol{w}_O(m) + v(m) \\ &= \boldsymbol{d}^T(m)\hat{\alpha}(m) - \mathbf{v}^T(m)\hat{\alpha}(m) + \overline{\mathbf{x}}^T(m)\boldsymbol{w}_O(m) + v(m) \end{aligned} \tag{3.9}$$

where $\boldsymbol{d}(m) = [d(m-1), \ldots, d(m-M)]^T$ and $\mathbf{v}(m) = [v(m-1), \ldots, v(m-M)]^T$. Define the decorrelated observation signal $\overline{d}(m)$ as

$$\overline{d}(m) \triangleq d_l - \boldsymbol{d}^T(m)\hat{\alpha}(m) \tag{3.10}$$

Thus, the decorrelated observation model is stated as

$$\overline{d}(m) = \overline{\mathbf{x}}^T(m)\boldsymbol{w}_o(m) + \overline{v}(m) \tag{3.11}$$

where $\overline{v}(m) = v(m) - \mathbf{v}^T(m)\hat{\alpha}(m)$.

The correntropy is a similarity measure between random variables $\mathcal{A}$ and $\mathcal{B}$, [12, 14], defined as

$$V(\mathcal{A}, \mathcal{B}) = E[k_\sigma(\mathcal{A}, \mathcal{B})] \tag{3.12}$$

where $E[\cdot]$ is the expectation operator and $k_\sigma(\mathcal{A}, \mathcal{B})$ is the kernel function with width σ. The kernel function is usually a Gaussian kernel, i.e.,

$$k_\sigma(\mathcal{A}, \mathcal{B}) = \exp\left(-\frac{(\mathcal{A}-\mathcal{B})^2}{2\sigma^2}\right) \tag{3.13}$$

Using the parzen estimation method, the correntropy estimation based on N samples is expressed as

$$\hat{V}(\mathcal{A}, \mathcal{B}) = \frac{1}{N} \sum_{i=1}^{N} k_{\sigma}(\mathcal{A}, \mathcal{B}) \tag{3.14}$$

For the decorrelated observation model presented in Eq. (3.11), the cost function is defined by using both the MCC criterion and weighted least squares (WLS) method as follows [15]:

$$J = k_{\sigma_1}\left(\|\overline{e}(m)\|^2_{r_{\overline{e}}^{-1}}\right) + k_{\sigma_1}\left(\|\mathbf{e}_{\omega}(m)\|^2_{\mathbf{P}_{\omega}^{-1}(m)}\right) \tag{3.15}$$

where $\overline{e}(m) = \overline{d}(m) - \overline{\mathbf{x}}^T(m)\boldsymbol{w}(m)$, $r_{\overline{e}} = E[\overline{v}_l^2] = \sigma_{\overline{v}}^2$, $\mathbf{P}_w(m) = E[\mathbf{e}_w(m)\mathbf{e}_w^T(m)]$, and $\mathbf{e}_w(m) = \boldsymbol{w}(m) - \boldsymbol{w}(m-1)$. To derive the optimal estimate $\boldsymbol{w}(m)$, it is necessary for the gradient of J with respect to $\boldsymbol{w}(m)$ to be zero, which is expressed as follows:

$$(\mathbf{P}_w^{-1}(m) + L_w(m)r_{\overline{e}}^{-1}\overline{\mathbf{x}}(m)\overline{\mathbf{x}}^T(m))\boldsymbol{w}(m) = \mathbf{P}_w^{-1}(m)\boldsymbol{w}(m-1) + L_w(m)r_{\overline{e}}^{-1}\overline{\mathbf{x}}(m)\overline{d}(m) \tag{3.16}$$

where $L_w(m) = k_{\sigma_1}\left(\|\overline{e}(m)\|^2_{r_{\overline{e}}^{-1}}\right)/k_{\sigma_1}\left(\|\mathbf{e}_w(m)\|^2_{\mathbf{P}_w^{-1}(m)}\right)$. To obtain the optimal estimation $\boldsymbol{w}_o(m)$ by adding and subtracting the term $L_{\omega}(m)r_{\overline{e}}^{-1}\overline{\mathbf{x}}(m)\overline{\mathbf{x}}^T(m)\boldsymbol{w}(m-1)$ to the right-hand side of Eq. (3.16) yields

$$\begin{aligned}(\mathbf{P}_w^{-1}(m) + L_w(m)r_{\overline{e}}^{-1}\overline{\mathbf{x}}(m)\overline{\mathbf{x}}^T(m))\boldsymbol{w}(m) =& L_w(m)r_{\overline{e}}^{-1}\overline{\mathbf{x}}(m)(\overline{d}(m) - \overline{\mathbf{x}}^T(m)\boldsymbol{w}(m-1)) \\ &+ (\mathbf{P}_w^{-1}(m) + L_w(m)r_{\overline{e}}^{-1}\overline{\mathbf{x}}(m)\overline{\mathbf{x}}^T(m))\boldsymbol{w}(m-1)\end{aligned} \tag{3.17}$$

If the term $(\mathbf{P}_w^{-1}(m) + L_w(m)r_{\overline{e}}^{-1}\overline{\mathbf{x}}(m)\overline{\mathbf{x}}^T(m))$ is reversible, we have

$$\boldsymbol{w}(m) = \boldsymbol{w}(m-1) + \mathbf{K}(m)(\overline{d}(m) - \overline{\mathbf{x}}^T(m)\boldsymbol{w}(m-1)) \tag{3.18}$$

where $\mathbf{K}(m) = L_w(m)r_{\overline{e}}^{-1}(\mathbf{P}_w^{-1}(m) + L_w(m)r_{\overline{e}}^{-1}\overline{\mathbf{x}}(m)\overline{\mathbf{x}}^T(m))^{-1}\overline{\mathbf{x}}(m)$ denotes Kalman gain. To simplify the calculation of $\mathbf{K}(m)$, using the matrix inversion lemma [16, 17] with $\mathbf{A} = \mathbf{P}_{\omega}^{-1}(m)$, $\mathbf{B} = \overline{\mathbf{x}}(m)$, $\mathbf{C} = L_w(m)r_{\overline{e}}^{-1}\mathbf{I}_L$ and $\mathbf{D} = \overline{\mathbf{x}}^T(m)$ yields

$$\begin{aligned}\mathbf{K}(m) &= L_w(m)r_{\overline{e}}^{-1}\left(\mathbf{P}_w(m) - \mathbf{P}_w(m)\overline{\mathbf{x}}(m)\left(L_w^{-1}(m)r_{\overline{e}} + \overline{\mathbf{x}}^T(m)\mathbf{P}_w(m)\overline{\mathbf{x}}(m)\right)^{-1}\overline{\mathbf{x}}^T(m)\mathbf{P}_w(m)\right)\overline{\mathbf{x}}(m) \\ &= L_w(m)r_{\overline{e}}^{-1}\left(\mathbf{P}_w(m)\overline{\mathbf{x}}(m) - \frac{\mathbf{P}_w(m)\overline{\mathbf{x}}(m)\overline{\mathbf{x}}^T(m)\mathbf{P}_w(m)\overline{\mathbf{x}}(m)}{L_{w(m)}^{-1}r_{\overline{e}} + \overline{\mathbf{x}}^T(m)\mathbf{P}_w(m)\overline{\mathbf{x}}(m)}\right) \\ &= L_w(m)r_{\overline{e}}^{-1}\left(\frac{L_w^{-1}(m)r_{\overline{e}}\mathbf{P}_w(m)\overline{\mathbf{x}}(m) + \mathbf{P}_w(m)\overline{\mathbf{x}}(m)\overline{\mathbf{x}}^T(m)\mathbf{P}_w(m)\overline{\mathbf{x}}(m) - \mathbf{P}_w(m)\overline{\mathbf{x}}(m)\overline{\mathbf{x}}^T(m)\mathbf{P}_w(m)\overline{\mathbf{x}}(m)}{L_w^{-1}(m)r_{\overline{e}} + \overline{\mathbf{x}}^T(m)\mathbf{P}_w(m)\overline{\mathbf{x}}(m)}\right) \\ &= \frac{L_w(m)\mathbf{P}_w(m)\overline{\mathbf{x}}(m)}{r_{\overline{e}} + L_{w(m)}\overline{\mathbf{x}}^T(m)\mathbf{P}_w(m)\overline{\mathbf{x}}(m)}\end{aligned} \tag{3.19}$$

In addition, according to the Riccati equation, the weight error covariance can be updated with

$$\mathbf{P}_w(m) = [\mathbf{I}_L - \mathbf{K}(m)\overline{\mathbf{x}}^T(m)][\mathbf{P}_\omega(m-1) + \sigma_q^2\mathbf{I}_L] \tag{3.20}$$

To further enhance the computational efficiency of the DRMCC algorithm, we configure the covariance matrix $\mathbf{P}_w(m)$ as $\sigma_w^2(m)\mathbf{I}_L$, where $\sigma_w^2(m) = \mathrm{Tr}(\mathbf{P}_w(m))/L$ and $\mathrm{Tr}(\cdot)$ is the trace operation. Hence, the Kalman gain is expressed as

$$\mathbf{K}(m) = \xi_w(m)\overline{\mathbf{x}}(m) = L_w(m)\sigma_w^2(m)\left(L_w(m)\sigma_w^2(m)\|\overline{\mathbf{x}}(m)\|^2 + \sigma_{\overline{v}}^2\right)^{-1}\overline{\mathbf{x}}(m) \tag{3.21}$$

where $\xi_w(m) = L_w(m)\sigma_w^2(m)\left(L_w(m)\sigma_w^2(m)\|\overline{\mathbf{x}}(m)\|^2 + \sigma_{\overline{v}}^2\right)^{-1}$. Taking Eq. (3.21) into Eq. (3.20), performing tracking operation and dividing by L, results in

$$\sigma_w^2(m) = (1 - \xi_w(m)\|\overline{\mathbf{x}}(m)\|^2/L)(\sigma_w^2(m-1) + \sigma_q^2) \tag{3.22}$$

Therefore, the DRMCC algorithm is summarized as

$$\xi_w(m) = L_w(m)\sigma_w^2(m-1)\left(L_w(m)\sigma_\omega^2(m-1)\|\overline{\mathbf{x}}(m)\|^2 + \sigma_{\overline{v}}^2\right)^{-1} \tag{3.23}$$

$$\sigma_w^2(m) = (1 - \xi_w(m)\|\overline{\mathbf{x}}(m)\|^2/L)(\sigma_w^2(m-1) + \sigma_q^2) \tag{3.24}$$

$$\mathbf{w}(m) = \mathbf{w}(m-1) + \xi_w(m)(\overline{d}(m) - \overline{\mathbf{x}}^T(m)\mathbf{w}(m-1))\overline{\mathbf{x}}(m) \tag{3.25}$$

Finally, we need to estimate $\boldsymbol{\alpha}$ in the iterative progression. Applying the earlier derivation process, we can effortlessly arrive at an updated equation for estimating the decorrelation coefficient vector $\hat{\alpha}(m)$, which is outlined as

$$\xi_\alpha(m) = L_\alpha(m)\sigma_\alpha^2(m-1)\left(L_\alpha(m)\sigma_\alpha^2(m-1)\|\mathbf{x}_\alpha(m)\|^2 + \sigma_z^2\right)^{-1} \tag{3.26}$$

Algorithm : 3.1 Pseudo code for DRMCC algorithm

Input: Inputs $x(m)$ and outputs $y(m)$ of the model
Output: Weight vector $w(m)$ of the filter
/* Initialization */
$m \leftarrow 1, \hat{\alpha}(-1) \leftarrow \mathbf{0}, \hat{\alpha}(0) \leftarrow \mathbf{0}, w(-1) \leftarrow \mathbf{0}, w(0) \leftarrow \mathbf{0}, \sigma_{\alpha}^2(0) \leftarrow 1, \sigma_w(m) \leftarrow 1$
/* Parameter settings */
Set kernel bandwidths σ_1, σ_2 . $\sigma_{\bar{v}}^2$, σ_q^2, and σ_z^2 are known or estimated online
/* Computation */
while $l \leq S$ **do**
$x_{\alpha}(m) \leftarrow [x(m-1), \ldots, x(l-M)]^T$
$x(m) \leftarrow [x(m), \ldots, x(m-L+1)]^T$
$X(m) \leftarrow [x(m-1), \ldots, x(m-M)]$
$y(m) \leftarrow [y(m-1), \ldots, y(m-M)]^T$
$e_x(m) \leftarrow x(m) - x_{\alpha}^T(m)\hat{\alpha}(m-1)$
$e_{\alpha}(m) \leftarrow \hat{\alpha}(m-1) - \hat{\alpha}(m-2)$
$L_{\alpha}(m) \leftarrow k_{\sigma_2}\left(\|e_x(m)\|_{\sigma_z^{-2}}^2\right) / k_{\sigma_2}\left(\|e_{\alpha}(m)\|_{\sigma_{\alpha}^{-2}(m-1)}^2\right)$
$\xi_{\alpha}(m) \leftarrow L_{\alpha}(m)\sigma_{\alpha}^2(m-1)\left(L_{\alpha}(m)\sigma_{\alpha}^2(m-1)\|x_{\alpha}(m)\|^2 + \sigma_z^2\right)^{-1}$
$\sigma_{\alpha}^2(m) \leftarrow \left[1 - \xi_{\alpha}(m)\|x_{\alpha}(m)\|^2 / M\right]\sigma_{\alpha}^2(m-1)$
$\hat{\alpha}(m) \leftarrow \hat{\alpha}(m-1) + \xi_{\alpha}(m)e_x(m)x_{\alpha}(m)$
$\bar{x}(m) \leftarrow x(m) - X(m)\hat{\alpha}(m)$
$\bar{y}(m) \leftarrow y(m) - y^T(m)\hat{\alpha}(m)$
$\bar{e}(m) \leftarrow \bar{y}(m) - \bar{x}^T(m)w(m-1)$
$e_w(m) \leftarrow w(m-1) - w(m-2)$
$L_w(m) \leftarrow k_{\sigma_1}\left(\|\bar{e}(m)\|_{\sigma_{\bar{v}}^{-2}}^2\right) / k_{\sigma_1}\left(\|e_w(m)\|_{\sigma_w^{-2}(m-1)}^2\right)$
$\xi_w(m) \leftarrow L_w(m)\sigma_w^2(m-1)\left(L_w(m)\sigma_w^2(m-1)\|\bar{x}(m)\|^2 + \sigma_{\bar{v}}^2\right)^{-1}$
$\sigma_w^2(m) \leftarrow \left[1 - \xi_w(m)\|\bar{x}(m)\|^2 / L\right][\sigma_w^2(m) + \sigma_q^2]$
$w(m) \leftarrow w(m-1) + \xi_w(m)\bar{e}(m)\bar{x}(m)$
$m \leftarrow m+1$
end
return w

$$\sigma_{\alpha}^2(m) = \left[1 - \xi_{\alpha}(m)\|\mathbf{x}_{\alpha}(m)\|^2/M\right]\sigma_{\alpha}^2(m-1) \tag{3.27}$$

$$\hat{\alpha}(m) = \hat{\alpha}(m-1) + \xi_{\alpha}(m)(x(m) - \mathbf{x}_{\alpha}^T(m)\hat{\alpha}(m-1))\mathbf{x}_{\alpha}(m) \tag{3.28}$$

where $L_{\alpha}(m) = k_{\sigma_2}\left(\|e_x(m)\|_{r_x^{-1}}^2\right)/k_{\sigma_2}\left(\|\mathbf{e}_{\alpha}(m)\|_{\mathbf{P}_{\alpha}^{-1}(m)}^2\right)$, $e_x(m) = x(m) - \mathbf{x}_{\alpha}^T(m)\hat{\alpha}(m)$, $r_x = E[z^2(m)] = \sigma_z^2$, $\mathbf{e}_{\alpha}(m) = \hat{\alpha}(m) - \hat{\alpha}(m-1)$, and $\mathbf{P}_{\alpha}(m) = E[\mathbf{e}_{\alpha}(m)\mathbf{e}_{\alpha}^T(m)]$.

Remark 3.1 It can be seen from Eqs. (3.23) to (3.28) that the DRMCC algorithm will degrade to the BDLMS algorithm when the kernel widths σ_1 and σ_2 tend to infinity.

Remark 3.2 As shown in the iterative equations of the DRMCC algorithm, the DRMCC algorithm can be regarded as a variable step-size decorrelation algorithm. Therefore, the proposed algorithm is effective in avoiding the tradeoff caused by constant step size.

Derived from the above analysis, Algorithm 1 provides the pseudocode for the DRMCC algorithm.

3.3 Performance Analysis

In this section, we focus on the mean stability and mean-square performance of the DRMCC algorithm from a statistical perspective. For subsequent analysis, we define the weight-error vector as

$$\tilde{\boldsymbol{w}}(m) = \boldsymbol{w}_o(m) - \boldsymbol{w}(m) \tag{3.29}$$

The errors $\overline{e}(m)$ and $\mathbf{e}_w(m)$ can be written as

$$\overline{e}(m) = \overline{d}(m) - \overline{\mathbf{x}}^T(m)\boldsymbol{w}(m-1) = \overline{\mathbf{x}}^T(m)\tilde{\boldsymbol{w}}(m-1) + \overline{\mathbf{x}}^T(m)\mathbf{q}(m) + \overline{v}(m) \tag{3.30}$$

$$\mathbf{e}_w = \boldsymbol{w}(m) - \boldsymbol{w}(m-1) = \tilde{\boldsymbol{w}}(m-1) + \mathbf{q}(m) - \tilde{\boldsymbol{w}}(m) \tag{3.31}$$

Therefore, the weight-error update equation is calculated by

$$\begin{aligned}\tilde{\boldsymbol{w}}(m) &= \tilde{\boldsymbol{w}}(m-1) + \mathbf{q}(m) - \xi_\omega(m)[\overline{\mathbf{x}}(m)\overline{\mathbf{x}}^T(m)\tilde{\boldsymbol{w}}(m-1) + \overline{\mathbf{x}}(m)\overline{\mathbf{x}}^T(m)\mathbf{q}(m) + \overline{\mathbf{x}}(m)\overline{v}(m)]\\ &= \mathbf{A}(m)\tilde{\boldsymbol{w}}(m-1) + \mathbf{A}(m)\mathbf{q}(m) - \xi_w(m)\overline{\mathbf{x}}(m)\overline{v}(m)\end{aligned} \tag{3.32}$$

where $\mathbf{A}(m) = \mathbf{I}_L - \xi_w(m)\overline{\mathbf{x}}(m)\overline{\mathbf{x}}^T(m)$. To perform performance analysis, we adopt the following typical assumptions.

(1) The noise $z(m)$ is a zero-mean random variable with variance $E[z^2(m)] = \sigma_z^2$.
(2) The noise $v(m)$ is an independent, identically distributed process with zero mean and variance $E\left[v^2(m)\right] = \sigma_v^2$ [9].
(3) The perturbation $\mathbf{q}(m)$ is a zero-mean random vector with covariance $\mathbf{Q} = E[\mathbf{q}(m)\mathbf{q}^T(m)] = \sigma_q^2\mathbf{I}_L$ [18].
(4) The decorrelated input vector $\overline{\mathbf{x}}(m)$, the gain $\xi_w(m)$, the noise $v(m)$, and the perturbation $\mathbf{q}(m)$ are mutually independent.
(5) L is long enough to ensure that the fluctuations of $\overline{\mathbf{x}}(m)$ at adjacent time are negligible.

Substituting Eq. (3.23) into Eq. (3.24) yields

$$\sigma_w^2(m) = \left\{1 - \frac{1}{L}\frac{L_w(m)\sigma_w^2(m-1)\|\overline{\mathbf{x}}(m)\|^2}{L_\omega(m)\sigma_\omega^2(m-1)\|\overline{\mathbf{x}}(m)\|^2 + \sigma_{\overline{v}}^2}\right\}[\sigma_w^2(m-1) + \sigma_q^2] \tag{3.33}$$

Using Assumptions 1 and 5, for sufficiently long L, $\sigma_w^2(m-1)$ is sufficiently small such that $E\big[\sigma_w^2(m)\|\overline{\mathbf{x}}(m)\|^2\big] \approx E\big[\sigma_w^2(m)\big]E\big[\|\overline{\mathbf{x}}(m)\|^2\big]$. And the expectation of gain $\xi_w(m)$ can be written as

$$E[\xi_w(m)] = \frac{E[L_w(m)]E\big[\sigma_w^2(m-1)\big]}{E[L_w(m)]E\big[\sigma_w^2(m-1)\big]E\big[\|\overline{\mathbf{x}}(m)\|^2\big] + \sigma_{\overline{v}}^2} \tag{3.34}$$

From Eq. (3.34), it is clear that the variance of $\xi_w(m)$ is also remarkably small. Consequently, $E[\sigma_w^2(m)]$ is approximated as

$$E[\sigma_w^2(m)] \approx \big[1 - E[\xi_\omega(m)]E\big[\|\overline{\mathbf{x}}(m)\|^2\big]/L\big]\Big\{E\big[\sigma_w^2(m-1)\big] + \sigma_q^2\Big\} \tag{3.35}$$

In addition, using the Taylor approximation, $L_w(m)$ can be expressed as

$$L_w(m) \approx \left\{1 - \frac{\|\overline{e}(m)\|^2_{\sigma_{\overline{v}}^{-2}}}{2\sigma_1^2}\right\} \Big/ \left\{1 - \frac{\|\mathbf{e}_w(m)\|^2_{\sigma_w^{-2}(m-1)}}{2\sigma_1^2}\right\} \tag{3.36}$$

Since L is sufficiently long, we can assume that $E\Big[\big\|\overline{\mathbf{x}}^T(m)\tilde{\mathbf{w}}(m-1)\big\|^2\Big]$ and $E\big[\|\tilde{\mathbf{w}}\|^2\big]$ is equal to zero. Therefore, the expectation of $L_w(m)$ is rewritten as

$$E[L_w(m)] \approx \left\{1 - \frac{\sigma_q^2\mathrm{Tr}(\mathbf{R}_{\overline{x}\overline{x}})+\sigma_{\overline{v}}^2}{2\sigma_1^2\sigma_{\overline{v}}^2}\right\} \Big/ \left\{1 - \frac{L\sigma_q^2}{2\sigma_1^2\sigma_w^2(m-1)}\right\} \tag{3.37}$$

where $\mathbf{R}_{\overline{x}\overline{x}}$ denotes the covariance matrix of the decorrelated input vector $\overline{\mathbf{x}}(m)$. Since is decorrelated, we can assume that $\mathbf{R}_{\overline{x}\overline{x}} = \sigma_z^2\mathbf{I}_L$, which implies that the decorrelated input signals are independent of each other and the covariance matrix of $\overline{\mathbf{x}}(m)$ is only related to the noise $z(m)$. Therefore, Eq. (3.37) is rewritten as

$$E[L_w(m)] \approx \left\{1 - \frac{L\sigma_q^2\sigma_z^2 + \sigma_{\overline{v}}^2}{2\sigma_1^2\sigma_{\overline{v}}^2}\right\} \Big/ \left\{1 - \frac{L\sigma_q^2}{2\sigma_1^2\sigma_w^2(m-1)}\right\} \tag{3.38}$$

According to the definition $\overline{v}(m) = v(m) - \mathbf{v}^T(m)\hat{\alpha}(m)$, the variance of $\overline{v}(m)$ can be written as

$$\sigma_{\overline{v}}^2 = \sigma_v^2\{1 + E\big[\|\hat{\alpha}(m)\|^2\big]\} \tag{3.39}$$

For analytical convenience, we directly utilize the expectation of $E\left[\|\hat{\alpha}(m)\|^2\right]$ obtained through ensemble averages.

Using the above assumptions and approximations, we can obtain the expectations of all variables in Eq. (3.34), which facilitates our subsequent analysis.

3.3.1 Mean Behavior

Using the assumptions 1–5, computing the expectation of Eq. (3.32) yields

$$E[\tilde{\boldsymbol{w}}(m)] = E[\mathbf{A}(m)]E[\tilde{\boldsymbol{w}}(m-1)] \tag{3.40}$$

where $E[\mathbf{A}(m)] = \mathbf{I}_L - E[\xi_w(m)]\mathbf{R}_{\overline{xx}} = [1 - E[\xi_w(m)]\sigma_z^2]\mathbf{I}_L$. Based on Eq. (3.40), the DRMCC algorithm converges asymptotically for any initial condition if the gain following conditions are satisfied

$$0 < E[\xi_w(m)] < \frac{2}{\lambda_{\max}(\mathbf{R}_{\overline{xx}})} \tag{3.41}$$

where $\lambda_{\max}(\cdot)$ indicates the maximum eigenvalue of the matrix.

3.3.2 Mean-Square Behavior

Firstly, the mean-square deviation (MSD) is described as

$$\mathrm{MSD}(m) = E\left[\|\tilde{\boldsymbol{w}}(m)\|^2\right] \tag{3.42}$$

Calculating the expectation of the squared weighted Euclidean norm on both sides of Eq. (3.32) and adopting Assumptions 1–5 yields

$$E\left[\|\tilde{\boldsymbol{w}}(m)\|^2\right] = E\left[\|\tilde{\boldsymbol{w}}(m-1)\|^2_{\mathbf{B}(m)}\right] + \mathrm{Tr}(\mathbf{B}(m)\mathbf{Q}) + \mathrm{Tr}(\sigma_{\overline{v}}^2 E[\xi_w(m)]^2 \sigma_z^2 \mathbf{I}_L) \tag{3.43}$$

where $\mathbf{B}(m) = E[\mathbf{A}^T(m)\mathbf{A}(m)]$. Using the vectorization operation, Eq. (3.43) is converted to

$$\begin{aligned} E\left[\|\tilde{\boldsymbol{w}}(m)\|^2_{\mathrm{vec}(\mathbf{I}_L)}\right] &= E\left[\|\tilde{\boldsymbol{w}}(m-1)\|^2_{\mathbf{F}(m)\mathrm{vec}(\mathbf{I}_L)}\right] + \mathrm{vec}(\mathbf{Q}^T)^T\mathbf{F}(m)\mathrm{vec}(\mathbf{I}_L) \\ &\quad + \sigma_{\overline{v}}^2\sigma_z^2 E[\xi_w(m)]^2 \mathrm{vec}(\mathbf{I}_L)^T \mathrm{vec}(\mathbf{I}_L) \end{aligned} \tag{3.44}$$

where $\mathbf{F}(m) = E[\mathbf{A}^T(m)] \otimes E[\mathbf{A}^T(m)]$, $\otimes$ denotes the Kronecker product, and $\mathrm{vec}(\cdot)$ represents the vectorization operator. Let $m = 1$, then Eq. (3.44) is written as

$$E\left[\|\tilde{\mathbf{w}}(1)\|^2_{\mathrm{vec}(\mathbf{I}_L)}\right] = E\left[\|\tilde{\mathbf{w}}(0)\|^2_{\mathbf{F}(1)\mathrm{vec}(\mathbf{I}_L)}\right] + \mathrm{vec}(\mathbf{Q}^T)^T\mathbf{F}(1)\mathrm{vec}(\mathbf{I}_L) + \sigma_{\bar{v}}^2\sigma_z^2 E[\xi_w(1)]^2\mathrm{vec}(\mathbf{I}_L)^T\mathrm{vec}(\mathbf{I}_L) \tag{3.45}$$

When $m = 2$, then Eq. (3.44) is expressed as

$$\begin{aligned}E\left[\|\tilde{\mathbf{w}}(2)\|^2_{\mathrm{vec}(\mathbf{I}_L)}\right] =& E\left[\|\tilde{\mathbf{w}}(1)\|^2_{\mathbf{F}(2)\mathrm{vec}(\mathbf{I}_L)}\right] + \mathrm{vec}(\mathbf{Q}^T)^T\mathbf{F}(2)\mathrm{vec}(\mathbf{I}_L)\\ &+ \sigma_{\bar{v}}^2\sigma_z^2 E[\xi_w(2)]^2\mathrm{vec}(\mathbf{I}_L)^T\mathrm{vec}(\mathbf{I}_L) = E\left[\|\tilde{\mathbf{w}}(0)\|^2_{\mathbf{F}(1)\mathbf{F}(2)\mathrm{vec}(\mathbf{I}_L)}\right]\\ &+ \mathrm{vec}(\mathbf{Q}^T)^T\mathbf{F}(1)\mathbf{F}(2)\mathrm{vec}(\mathbf{I}_L) + \sigma_{\bar{v}}^2\sigma_z^2 E[\xi_w(1)]^2\mathrm{vec}(\mathbf{I}_L)^T\mathbf{F}(2)\mathrm{vec}(\mathbf{I}_L)\\ &+ \mathrm{vec}(\mathbf{Q}^T)^T\mathbf{F}(2)\mathrm{vec}(\mathbf{I}_L) + \sigma_{\bar{v}}^2\sigma_z^2 E[\xi_w(2)]^2\mathrm{vec}(\mathbf{I}_L)^T\mathrm{vec}(\mathbf{I}_L)\\ &= E\left[\|\tilde{\mathbf{w}}(0)\|^2_{\left(\prod_{i=1}^{2}\mathbf{F}(i)\right)\mathrm{vec}(\mathbf{I}_L)}\right] + \mathrm{vec}(\mathbf{Q}^T)^T\left[\left(\prod_{j=i}^{2}\mathbf{F}(i)\right) + \mathbf{F}(2)\right]\mathrm{vec}(\mathbf{I}_L)\\ &+ \sigma_{\bar{v}}^2\sigma_z^2\mathrm{vec}(\mathbf{I}_L)^T\left\{E[\xi_w(1)]^2\mathbf{F}(2) + E[\xi_w(2)]^2\right\}\mathrm{vec}(\mathbf{I}_L)\end{aligned} \tag{3.46}$$

Therefore, iterating Eq. (3.44), we have

$$\begin{aligned}E\left[\|\tilde{\mathbf{w}}(0)\|^2_{\mathrm{vec}(\mathbf{I}_L)}\right] =& E\left[\|\tilde{\mathbf{w}}_0\|^2_{\left(\prod_{i=1}^{l}\mathbf{F}(i)\right)\mathrm{vec}(\mathbf{I}_L)}\right] + \mathrm{vec}(\mathbf{Q}^T)^T\left[\sum_{i=1}^{l}\left(\prod_{j=i}^{l}\mathbf{F}(i)\right)\right]\mathrm{vec}(\mathbf{I}_L)\\ &+ \sigma_{\bar{v}}^2\sigma_z^2\mathrm{vec}(\mathbf{I}_L)^T\left\{\sum_{i=1}^{l-1}E[\xi_w(i)]^2\left(\prod_{j=i+1}^{l}\mathbf{F}(j)\right)\right\}\mathrm{vec}(\mathbf{I}_L)\\ &+ \sigma_{\bar{v}}^2\sigma_z^2 E[\xi_w(m)]^2\mathrm{vec}(\mathbf{I}_L)^T\mathrm{vec}(\mathbf{I}_L)\end{aligned} \tag{3.47}$$

Subtracting Eq. (3.47) at time m from Eq. (3.47) at time $m - 1$ yields

$$\begin{aligned}E\left[\|\tilde{\mathbf{w}}(m)\|^2_{\mathrm{vec}(\mathbf{I}_L)}\right] =& E\left[\|\tilde{\mathbf{w}}(m-1)\|^2_{\mathrm{vec}(\mathbf{I}_L)}\right] + \sigma_{\bar{v}}^2\sigma_z^2 E[\xi_w(m)]^2\mathrm{vec}(\mathbf{I}_L)^T\mathrm{vec}(\mathbf{I}_L)\\ &- \sigma_{\bar{v}}^2\sigma_z^2 E[\xi_w(m-1)]^2\mathrm{vec}(\mathbf{I}_L)^T\mathrm{vec}(\mathbf{I}_L)\\ &+ E\left[\|\tilde{\mathbf{w}}(0)\|^2_{\left(\prod_{i=1}^{l}\mathbf{F}(i) - \prod_{i=1}^{l-1}\mathbf{F}(i)\right)\mathrm{vec}(\mathbf{I}_L)}\right]\\ &+ \mathrm{vec}(\mathbf{Q}^T)^T\left[\mathbf{F}(m) + \sum_{i=1}^{l-1}\left(\prod_{j=i}^{l-1}\mathbf{F}(j)\right)\mathbf{F}(m) - \sum_{i=1}^{l-1}\left(\prod_{j=i}^{l-1}\mathbf{F}(j)\right)\right]\mathrm{vec}(\mathbf{I}_L)\\ &+ \sigma_{\bar{v}}^2\sigma_z^2\mathrm{vec}(\mathbf{I}_L)^T\left\{E[\xi_w(m-1)]^2\mathbf{F}(m) + \sum_{i=1}^{l-2}E[\xi_w(i)]^2\left(\prod_{j=i+1}^{l-1}\mathbf{F}(j)\right)\mathbf{F}(m)\right.\end{aligned}$$

$$-\sum_{i=1}^{l-2} E[\xi_w(i)]^2\left(\prod_{j=i+1}^{l-1} \mathbf{F}(j)\right)\Bigg\}\text{vec}(\mathbf{I}_L = E\left[\|\tilde{\mathbf{w}}(m-1)\|^2_{\text{vec}(\mathbf{I}_L)}\right]$$

$$+\sigma_{\bar{v}}^2\sigma_z^2 E[\xi_\omega(m)]^2\text{vec}(\mathbf{I}_L)^T\text{vec}(\mathbf{I}_L) + E\left[\|\tilde{\mathbf{w}}_0\|^2_{\left(\prod_{i=1}^{l-1}\mathbf{F}(i)\right)(\mathbf{F}(m)-\mathbf{I}_{L^2})\text{vec}(\mathbf{I}_L)}\right]$$

$$+\text{vec}(\mathbf{Q}^T)^T\left[\sum_{i=1}^{l-1}\left(\prod_{j=i}^{l-1}\mathbf{F}(j)\right)\right](\mathbf{F}(m)-\mathbf{I}_{L^2})\text{vec}(\mathbf{I}_L) + \text{vec}(\mathbf{Q}^T)^T\mathbf{F}(m)\text{vec}(\mathbf{I}_L)$$

$$+\sigma_{\bar{v}}^2\sigma_z^2\text{vec}(\mathbf{I}_L)^T\left\{\sum_{i=1}^{l-2}E[\xi_\omega(i)]^2\left(\prod_{j=i+1}^{l-1}\mathbf{F}(j)\right)\right\}(F(m)-I_{L^2})\text{vec}(\mathbf{I}_L)$$

$$+\sigma_{\bar{v}}^2\sigma_z^2 E[\xi_w(m-1)]^2\text{vec}(\mathbf{I}_L)^T(\mathbf{F}(m)-\mathbf{I}_{L^2})\text{vec}(\mathbf{I}_L) \quad (3.48)$$

Utilizing Eq. (3.48), it becomes straightforward to derive the transient MSD behavior of the DRMCC algorithm.

3.3.3 Complexity Analysis

In this subsection, we compare the computational complexity of the DRMCC algorithm with several existing algorithms. The number of addition, multiplication and exponentiation operations per iteration are summarized in Table 3.1, where P denotes the projection order.

As can be seen from Table 3.1, the computational complexity of all the algorithms increases with the filter length L. In addition, the computational complexity of the APMCC algorithm exhibits a cubic relationship with the projection order and has the highest computational complexity among all algorithms. The computational complexity of the DRMCC algorithm is slightly higher than the MCC algorithm and the NMCC algorithm.

Table 3.1 Comparison of computational complexity

Algorithms	Addition	Multiplication	Exponentiation
MCC	$2L$	$2L+6$	1
NMCC	$2L+1$	$3L+7$	1
APMCC	$P^3+2LP^2-P^2+LP-P+L$	$P^3+2LP^2+2P^2+LP+4P$	P
DRMCC	$LM+7M+6L-1$	$LM+6M+5L+36$	4

3.3.4 Parameter Estimation

Given that $\sigma_{\overline{v}}^2$, σ_q^2, and σ_z^2 are generally unknown in real-world applications, it becomes necessary to undertake their estimation during operation.

Substituting Eq. (3.11) into Eq. (3.25) yields

$$\mathbf{w}(m) = \mathbf{w}(m-1) + \xi_w(m)\overline{\mathbf{x}}(m)\overline{v}(m) + \xi_w(m)\overline{\mathbf{x}}(m)\overline{\mathbf{x}}^T(m)(\mathbf{w}_o(m) - \mathbf{w}(m-1)) \tag{3.49}$$

Subtracting both sides of Eq. (49) from $\mathbf{w}_o(m)$ and pre-multiplying $\overline{\mathbf{x}}(m)$ on both sides of the result yields

$$\begin{aligned}\overline{\mathbf{x}}^T(m)(\mathbf{w}_o(m) - \mathbf{w}(m)) &= \left(1 - \xi_\omega(m)\|\overline{\mathbf{x}}(m)\|^2\right)\overline{\mathbf{x}}^T(m)(\mathbf{w}_o(m) - \mathbf{w}(m-1)) \\ &\quad - \xi_\omega(m)\|\overline{\mathbf{x}}(m)\|^2\overline{v}\end{aligned} \tag{3.50}$$

When close to the steady state in the stationary scenario ($\sigma_q^2 = 0$), $\sigma_w^2(m-1)$ is small enough so that $\xi_w(m) \approx 0$. And in the nonstationary scenario, the parameter estimation of $\sigma_{\overline{v}}^2(m)$ will be discussed in the estimation of σ_q^2. To eliminate the unknown weight vector $\mathbf{w}_o(m)$ in Eq. (3.50), we assume that $\mu(m) = \xi_w(m)\|\overline{\mathbf{x}}(m)\|^2$ is small enough. Furthermore, since the performance of the algorithm is determined by the ratio between $\sigma_{\overline{v}}^2(m)$ and $\sigma_z^2(m)$ [19], Eq. (3.50) can be rewritten as

$$\begin{aligned}\mu(m)\overline{v} &\approx \overline{\mathbf{x}}^T(m)(\mathbf{w}_o(m) - \mathbf{w}(m-1)) - \overline{\mathbf{x}}^T(m)(\mathbf{w}_o(m) - \mathbf{w}(m)) \\ &= \overline{\mathbf{x}}^T(m)(\mathbf{w}(m) - \mathbf{w}(m-1))\end{aligned} \tag{3.51}$$

By computing the expectation of the squares of both sides of Eq. (3.51) [20], we derive

$$\sigma_{\overline{v}}^2(m) = \frac{1}{\mu^2(m)} E\left[\left[\overline{\mathbf{x}}^T(m)(\mathbf{w}(m) - \mathbf{w}(m-1))\right]^2\right] \tag{3.52}$$

Using exponential moving average [21], $\sigma_{\overline{v}}^2(m)$ is represented as

$$\sigma_{\overline{v}}^2(m) = \lambda_{\overline{v}}\sigma_{\overline{v}}^2(m-1) + (1 - \lambda_{\overline{v}})\left[\overline{\mathbf{x}}^T(m)(\mathbf{w}(m) - \mathbf{w}(m-1))\right]^2 \tag{3.53}$$

where $\lambda_{\overline{v}}$ denotes the forgetting factor.

Substituting Eq. (3.3) into Eq. (3.28) yields

$$\hat{\alpha}(m) = \hat{\alpha}(m-1) + \xi_\alpha(m)\mathbf{x}_\alpha(m)z(m) + \xi_\alpha(m)\mathbf{x}_\alpha(m)\mathbf{x}_\alpha^T(m)(\boldsymbol{\alpha} - \hat{\alpha}(m-1)) \tag{3.54}$$

Similarly, an estimate of $\sigma_z^2(m)$ is written as

$$\sigma_z^2(m) = \lambda_z \sigma_z^2(m-1) + (1-\lambda_z)\left[\overline{\mathbf{x}}_\alpha^T(m)(\hat{\alpha}(m) - \hat{\alpha}(m))\right]^2 \tag{3.55}$$

where λ_z denotes the forgetting factor.

Based on Eq. (3.2), we arrive at

$$L\sigma_q^2 = E\left[\|\mathbf{w}_o(m) - \mathbf{w}_o(m-1)\|^2\right] \tag{3.56}$$

Given that $\mathbf{w}_o(m)$ and $\mathbf{w}_o(m-1)$ are unknown, we replace them with $\mathbf{w}(m)$ and $\mathbf{w}(m-1)$ from the iterative process, i.e.,

$$L\sigma_q^2 = E\left[\|\mathbf{w}(m) - \mathbf{w}(m-1)\|^2\right] \tag{3.57}$$

Therefore, σ_q^2 is given as

$$\sigma_w^2(m) = \lambda_q \sigma_w^2(m-1) + (1-\lambda_q)\|\mathbf{w}(m) - \mathbf{w}(m-1)\|^2 \tag{3.58}$$

$$\sigma_q^2(m) = \sigma_w^2(m)/L \tag{3.59}$$

where λ_q represents the forgetting factor. When the filter length L is long enough in a nonstationary scenario, $\sigma_q^2(m)$ is small enough. Therefore, $\sigma_{\overline{v}}^2(m)$ can be similarly estimated by Eq. (3.53).

3.4 Computer Simulation Examples

In this section, Monte Carlo (MC) simulations are performed to corroborate the theoretical model under both stationary and nonstationary scenarios. In addition, we evaluate the learning behavior of the developed DRMCC contrasted to other considered algorithms under non-Gaussian environments. In the simulation, the unknown weight vector $\mathbf{w}_o(m)$ and the correlation indices $\boldsymbol{\alpha}$ are produced randomly. The correlated input signal $x(m)$ is produced according to Eq. (3.3), with both $z(m)$ and $\mathbf{q}(m)$ being zero-mean Gaussian sequence and $\sigma_z^2 = 1$. The filter's weight vector is null vector. Furthermore, the simulation curves are displayed via averaging results from 50 independent MC trials.

3.4.1 *Validation of Theoretical Model*

First, we give the evolution of MSDs for the DRMCC algorithm in the stationary and nonstationary scenarios with $L = 10$ and $M = 2$. The noise $v(m)$ obeys the binary distribution over $\{-p, 1-p\}$, where the probability mass function is defined as $Pr(v(m) = 1-p) = p$. Figure 3.1 gives the transient behavior of the MSDs

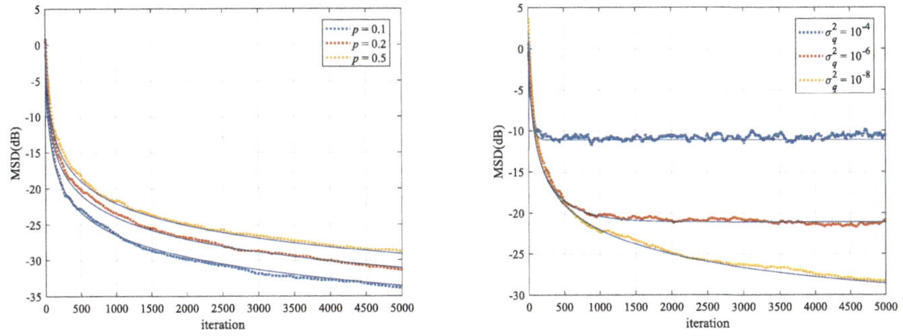

(a) (b)

Fig. 3.1 Simulated and theoretical learning curves under **a** the stationary and **b** the nonstationary scenarios

under the stationary and nonstationary scenarios, where the dotted line refers to the simulation results, and the solid lines depict transient behavior. As shown in Fig. 3.1, the theoretical curve resulted from Eq. (3.48) matches very well with the simulation results.

3.4.2 Performance Comparison in System Identification

Second, we assess the performance of the DRMCC algorithm against other algorithms in the stationary and nonstationary scenarios. The correlated input signal is produced by $x(m) = tx(m-1) + \sqrt{1-t^2}x(m-2) + z(m)$, where $t = 0.99$. The noise $v(m)$ is generated using the Bernoulli-Gaussian and alpha-stable models. The Bernoulli-Gaussian model is expressed as $v(m) = b(m)\mathcal{N}(0, 0.01) + (1 - b(m))\mathcal{N}(0, 10000)$, where $\mathcal{N}(0, \varepsilon^2)$ represents the zero-mean Gaussian distribution with variance ε^2, and $b(m)$ indicates the binary distribution over $\{0, 1\}$ with the probability mass function defined as $Pr(v(m) = 1) = 0.9$ [22]. The alpha-stable model is defined with tail parameter $\alpha = 1.2$ and dispersion parameter $\gamma = 0.1$. In the nonstationary scenario, the perturbation variance is set to $\sigma_q^2 = 10^{-7}$ for the Bernoulli-Gaussian environment and $\sigma_q^2 = 10^{-6}$ for the alpha-stable environment. The parameters of the competing algorithms are adjusted so that they have the same error in the nonstationary scenario. The used algorithm parameters are summarized in Table 3.2. The regularization parameters for the NLMS, NMCC, AP and APMCC algorithms are chosen as $\lambda = 20\sigma_x^2$, where σ_x^2 represents the variance of the input signal [6]. The projection order of AP and APMCC algorithms is 8.

The MSD convergence curves for the algorithms in various noise environments are illustrated in Figs. 3.2 and 3.3. As depicted in Figs. 3.2 and 3.3, the algorithms based on the MMSE criterion do not converge in these impulsive noise environments, while the algorithm based on the MCC criterion exhibits robust performance against impulsive noise. It is worth noting that the BDLMS algorithm performs

Table 3.2 Parameter selection for the used algorithm in system identification

Algorithms	Step size	Kernel width	λ_z	λ_q	$\lambda_{\overline{v}}$
LMS [2]	0.01	–	–	–	–
NLMS [23]	0.3	–			
AP [24]	0.18	–			
BDLMS [9]	–	–	0.9999	0.999999	0.999999
MCC [25]	0.01	1			
NMCC [26]	0.3	1			
APMCC [27]	0.18	1			
DRMCC	–	$\sigma_1 = 1, \sigma_2 = 4$	0.9999	0.999999	0.999999

worse in nonstationary scenario than the LMS, NLMS, and AP algorithms because the BDLMS algorithm in such scenario needs to estimate the parameter σ_q^2 associated with the filter weights as shown in Eq. (3.59). The weights of the BDLMS algorithm based on the MMSE criterion are susceptible to impulsive noise leading to inaccurate estimation of σ_q^2, which makes the algorithm perform worse than other algorithms.

Since the weights of the DRMCC algorithm are insensitive to the impulsive noise, it still has better performance than the BDLMS algorithm in nonstationary scenario. It can be seen from Fig. 3.2 that the DRMCC algorithm outperforms the MCC, NMCC and APMCC algorithms in terms of convergence speed and steady state error in the stationary scenario. As shown in Fig. 3.3, the DRMCC algorithm has a faster convergence rate in nonstationary scenario when the same steady state error is achieved than the MCC, NMCC and APMCC algorithms. In addition, the variable step size $\xi_w(m)$ of the DRMCC algorithm and the value of the step size in the MCC under stationary scenario in Bernoulli-Gaussian noise environment are shown in Fig. 3.4. As can be seen in Fig. 3.4, the step size is larger in the initial phase

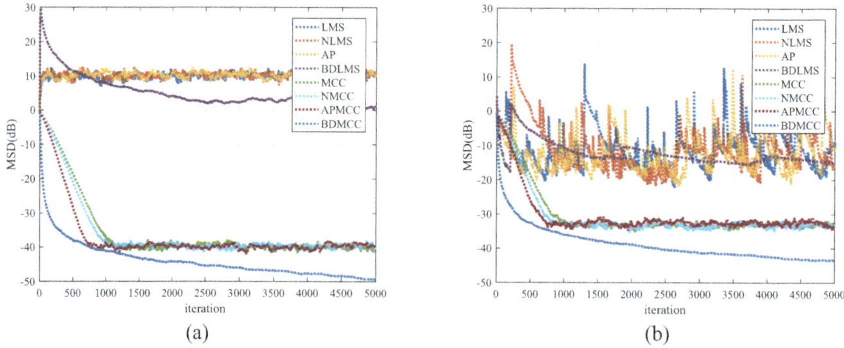

Fig. 3.2 Performance comparison under stationary scenario in **a** Bernoulli-Gaussian noise and **b** alpha-stable noise environments

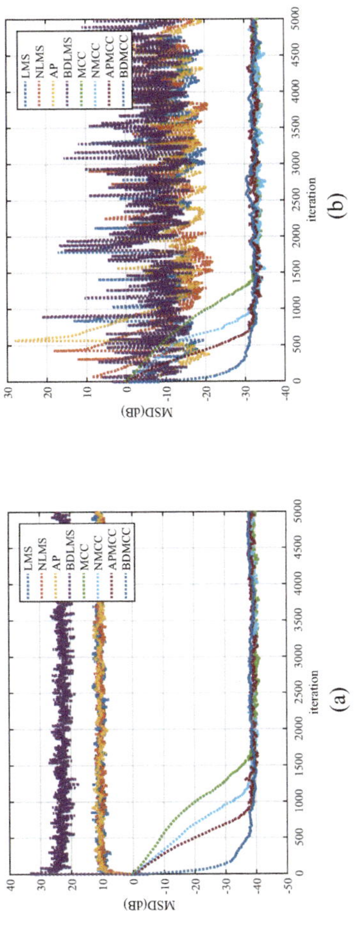

Fig. 3.3 Performance comparison under nonstationary scenario in **a** Bernoulli-Gaussian noise and **b** alpha-stable noise environments

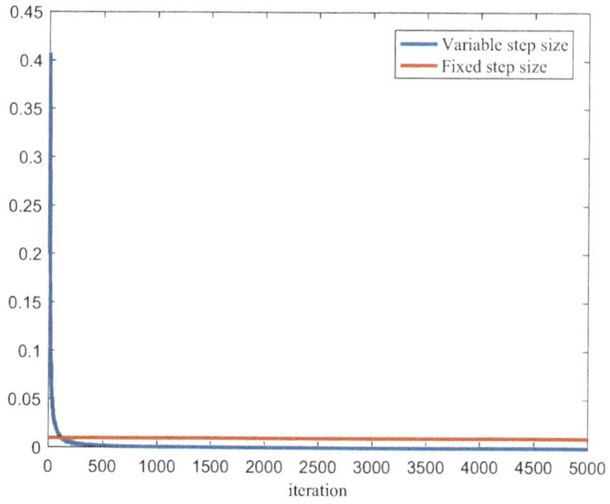

Fig. 3.4 Variable step-size curve under stationary scenario in Bernoulli-Gaussian noise environment

and smaller as the algorithm approaches the steady state. Therefore, the proposed algorithm is effective in avoiding the tradeoff caused by constant step size.

3.4.3 Acoustic Echo Cancellation

Finally, we evaluate the DRMCC algorithm's performance under both single-talk and double-talk situations in AEC. We use the common normalized MSD (NMSD) as the evaluation metric, which can be defined as

$$\mathrm{NMSD}(m) = 10\log_{10}E\left[\|\boldsymbol{w}(m)\|^2/\|\boldsymbol{w}_o\|^2\right] \tag{3.60}$$

The impulse response of the echo path $\boldsymbol{w}_o$ with length $L = 2048$ [6]. The sampling frequency and sampling time for both the far-end and near-end speech signals are 8 kHz and 20 s, respectively. To evaluate the tracking performance of the algorithms, we simulate abrupt changes in the echo path by right-shifting $\boldsymbol{w}_o$ by 12 bits during the iteration process. The used algorithm parameters are summarized in Table 3.3. The regularization parameters for the AP and APMCC algorithms are chosen as $\lambda = 20\sigma_x^2$. The projection order of APMCC algorithms is 4. The order of the autoregressive model is chosen to be 2. The noise $v(m)$ is a dual source, including the zero-mean Gaussian noise with variance of 0.01 and the alpha-stable distribution noise with tail parameter $\alpha = 1.2$ and dispersion parameter $\gamma = 0.01$. The NMSD curves for all the algorithms under both single-talk and double-talk scenarios in alpha-stable noise environments are given in Fig. 3.5. Figure 3.5 illustrates that the DRMCC algorithm achieves the quicker adaption speed and maintains the higher estimation accuracy contrasted to other algorithms. It is noteworthy that the DRMCC algorithm exhibits insensitivity to double-talk interference.

Table 3.3 Parameter selection for the used algorithm in AEC

Algorithms	Step size	Kernel width	λ_z	λ_q	$\lambda_{\overline{v}}$
MCC	0.00008	10			
NMCC	0.6	10			
APMCC	0.06	10			
DRMCC	–	$\sigma_1 = 10, \sigma_2 = 10$	0.999996	0.9999	0.999999

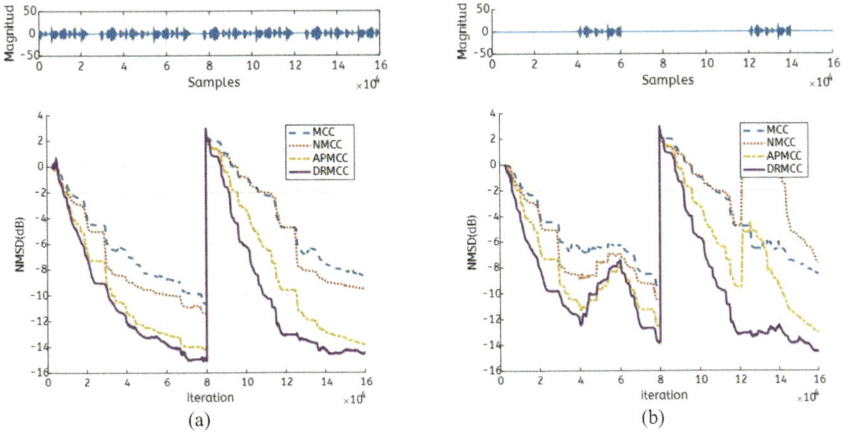

Fig. 3.5 Comparison of NMSD convergence curves of the algorithms under alpha-stable noise environment in **a** single-talk and **b** double-talk scenarios

3.5 Summary

This paper presents an observation model-based DRMCC algorithm for AEC in non-Gaussian noise environments. Furthermore, the corresponding convergence conditions and transient behavior in the mean-square sense are discussed. Finally, the parameter estimation method is given to solve the problem where the noise variance $\sigma_{\overline{v}}^2$, the perturbation variances σ_q^2 and σ_z^2 are unknown. Simulation results demonstrate a good agreement between the theoretical model and experimental results in a non-Gaussian noise environment. It is noteworthy that the DRMCC algorithm significantly enhance the convergence performance in non-Gaussian noise environments compared with existing adaptive filtering algorithms.

References

1. S. Haykin, *Adaptive Filter Theory*. (Prentice-Hall, Englewood Cliffs, NJ, USA, 2002)
2. A.H. Sayed, *Fundamentals of Adaptive Filtering*. (Wiley, Hoboken, NJ, USA, 2003)

3. J.J. Jeong, S.h. Kim, G. Koo, S.W. Kim, Mean-square deviation analysis of multiband-structured subband adaptive filter algorithm. IEEE Trans. Signal Process. **64**, 986–994 (2016)
4. H. Zhao, D. Liu, S. Lv, Robust maximum correntropy criterion subband adaptive filter algorithm for impulsive noise and noisy input. IEEE Trans. Circuits Syst. II Express Briefs **69**, 604–608 (2022)
5. J. Yoo, J. Shin, P. Park, Variable step-size affine projection sign algorithm. IEEE Trans. Circuits Syst. II Express Briefs **61**, 274–278 (2014)
6. S. Lv, H. Zhao, W. Xu, Robust widely linear affine projection m-estimate adaptive algorithm: Performance analysis and application. IEEE Trans. Signal Process. **71**, 3624–3636 (2023)
7. S. Zhang, H.C. So, W. Mi, H. Han, A family of adaptive decorrelation nlms algorithms and its diffusion version over adaptive networks. IEEE Trans. Circuits Syst. I Regul. Pap. **65**, 638–649 (2023)
8. S. Zhang, J. Zhang, H.C. So, Low-complexity decorrelation nlms algorithms: Performance analysis and aec application. IEEE Trans. Signal Process. **68**, 6621–6632 (2020)
9. F. Huang, S. Zhang, J. Zhang, H. Chen, A.H. Sayed, Diffusion Bayesian decorrelation algorithms over networks. IEEE Trans. Signal Process. **71**, 571–586 (2023)
10. R. Pogula, T.K. Kumar, F. Albu, Robust sparse normalized lmat algorithms for adaptive system identification under impulsive noise environments. Circuits, Syst., Signal Process. **38**, 5103C5134 (2019)
11. J.C. Principe, *Information Theoretic Learning: Renyi's Entropy And Kernel Perspectives.* (Springer, New York, USA, 2010)
12. H. Song, D. Ding, H. Dong, Q.-L. Han, Distributed maximum correntropy filtering for stochastic nonlinear systems under deception attacks. IEEE Trans. Cybern. **52**, 3734–3744 (2022)
13. Y. Wang, Y. Li, F. Albu, R. Yang, Group-constrained maximum correntropy criterion algorithms for estimating sparse mix-noised channels. Circuits Syst. Signal Process. **19**, 432 (2017)
14. W. Liu, P.P. Pokharel, J.C. Principe, Correntropy: properties and applications in non-gaussian signal processing. IEEE Trans. Signal Process. **55**, pp. 5287–5298 (2007)
15. R. Izanloo, S.A. Fakoorian, H.S. Yazdi, D. Simon, Kalman filtering based on the maximum correntropy criterion in the presence of non-gaussian noise, in *2016 Annual Conference on Information Science and Systems (CISS)* (2016), pp. 500–505
16. X. Hou, H. Zhao, X. Long, W. Jin, The kernel recursive maximum total correntropy algorithm. IEEE Trans. Circuits Syst. II Express Briefs **69**, 5139–5143 (2022)
17. Y. Engel, S. Mannor, R. Meir, The kernel recursive least-squares algorithm. IEEE Trans. Signal Process. **52**, 2276–2285 (2004)
18. J. Chen, C. Richard, A.H. Sayed, Diffusion lms over multitask networks. IEEE Trans. Signal Process. **63**, 2734–2748 (2015)
19. S. Akhlaghi, N. Zhou, Z. Huang, Adaptive adjustment of noise covariance in kalman filter for dynamic state estimation, in *2017 IEEE Power and Energy Society General Meeting* (2017), pp. 1–5
20. J.G. Proakis, *Digital Communications.* (Prentice-Hall, Englewood Cliffs, NJ, USA, 2001)
21. L. Shi, H. Zhao, W. Wang, L. Lu, Combined regularization parameter for normalized lms algorithm and its performance analysis. Signal Process. **162**, 76–82 (2019)
22. J. Rong, J. Zhang, H. Duan, Robust sparse bayesian learning based on the bernoulli-gaussian model of impulsive noise. Digit. Signal Process. **136**, 104013 (2023)
23. H.-C. Huang, J. Lee, A new variable step-size nlms algorithm and its performance analysis. IEEE Trans. Signal Process. **60**, 2056–2060 (2012)
24. H.-C. Shin, A. Sayed, Mean-square performance of a family of affine projection algorithms. IEEE Trans. Signal Process. **52**, 90–102 (2004)
25. X. Hou, H. Zhao, X. Long, An innovative transient analysis of adaptive filter with maximum correntropy criterion. IEEE Signal Process. Lett. **29**, 1689–1693 (2022)

26. W. Ma, D. Zheng, Y. Li, Z. Zhang, B. Chen, Bias-compensated normalized maximum correntropy criterion algorithm for system identification with noisy input. Signal Process. **152**, 160–164 (2018)
27. C. Qiu, G. Qian, S. Wang, Widely linear maximum complex correntropy criterion affine projection algorithm and its performance analysis. IEEE Trans. Signal Process. **70**, 3540–3550

Chapter 4
Combinatoric Correntropy Criterion

4.1 Introduction

In order to have a better performance of the adaptive filtering algorithms, it is often necessary to find a suitable cost function [1, 2]. Currently, it is known that designing the corresponding adaptive filtering algorithms, such as LMS and NLMS algorithms, using the mean-square error (MSE) criterion as the cost function, can allow the algorithms to have low computational complexity, smoothness, and convexity, among others [3, 4]. However, the performance of MSE-based algorithms degrades significantly in the presence of non-Gaussian noise, so it is crucial to find new cost functions that are resistant to the non-Gaussian noise that is widely present in the real environment [5].

But the treatment of the non-Gaussian noise interference is often a more difficult part within the field of signal processing. This problem has been solved by a Gaussian-like modeling approach based on non-quadratic cost functions, and the feasibility of this approach has been demonstrated [6].

In recent years, the study of the correntropy, a nonlinear similarity measure, has received a lot of attention from a wide range of researchers. Because the correntropy contains higher-order error information, it effectively improves the robustness of adaptive filtering algorithms. In addition, the adaptive filtering algorithms under the maximum correntropy criterion (MCC) have a similar computational complexity with the algorithms under the MSE criterion [7]. Therefore, many adaptive algorithms based on the MCC have been proposed by researchers to optimize the performance of traditional MSE-based algorithms in the non-Gaussian environment [8, 9].

In the existing work, many methods were proposed to improve the performance of traditional MCC algorithms, such as GMCC, MTC, the mixture MCC algorithm (MMCC), affine projection MCC algorithms, and so on [10–16]. However, all these methods start from a single kernel and do not consider the case of multiple kernels. To address the above problem, this chapter proposes a combinatoric correntropy and its adaptive algorithm. Compared to the traditional single-kernel correntropy algorithms,

H. Zhao et al., *Information Theoretic Learning-Based Filter*,
https://doi.org/10.1007/978-3-032-29623-8_4

the C-MCC algorithm has a more generalized form. In the special case, the C-MCC algorithm will reduce to the MCC algorithm. This chapter gives the corresponding properties and proofs of the C-C. In addition, the theoretical performance of the C-MCC algorithm is analyzed by using the Taylor expansion method.

4.2 Combinatoric Correntropy Criterion

The correntropy for two variables c_1 and c_2 is defined as:

$$V_{c,\sigma}(c_1, c_2) = E\left[\kappa_{c,\sigma}(c_1 - c_2)\right], \tag{4.1}$$

where c_1 and c_2 are real-valued variables, $\kappa_{c,\sigma}(c_1 - c_2)$ is the kernel function, σ is the kernel-width, and $E[\cdot]$ stands for expected operation. When the Gaussian kernel is used as kernel function, that is

$$\kappa_{c,\sigma}(c_1 - c_2) = G_{c,\sigma}(c_1 - c_2) = \frac{1}{2\pi\sigma^2}\exp\left(-\frac{(c_1 - c_2)^2}{2\sigma^2}\right). \tag{4.2}$$

In this chapter, the combined correntropy consists of two kinds of correntropy of a single kernel, i.e.,

$$V^M_{c,\sigma}(c_1, c_2) = E\left[\kappa^M_{c,\sigma}(c_1 - c_2)\right], \tag{4.3}$$

where

$$\kappa^M_{c,\sigma}(c_1 - c_2) = \exp\left[-\left(\lambda_1\frac{(c_1 - c_2)^2}{2\sigma_1^2} + \lambda_2\frac{|c_1 - c_2|}{\sigma_2^2}\right)\right], \tag{4.4}$$

and λ_1, λ_2 are the combined parameters that can determine the proportion of the two types of kernels and $\lambda_1 + \lambda_2 = 1$. When $\lambda_1 \to 1$ or $\lambda_2 \to 1$, $\kappa^M_{c,\sigma}(c_1 - c_2)$ will become a single kernel. To make a difference with the conventional correntropy, the correntropy criterion based on the kernel function $V^M_{c,\sigma}$ is called the combinatoric maximum correntropy criterion.

In practice, the joint distribution function is unknown for two random variables and the estimation is obtained via the finite samples $\{c_{1,i}, c_{2,i}\}_{i=1}^N$, in this case:

$$\begin{aligned}\hat{V}^M_{c,\sigma}(c_1, c_2) &= \frac{1}{N}\sum_{i=1}^N \exp\left[-\left(\lambda_1\frac{\left(c_{1,i} - c_{2,i}\right)^2}{2\sigma_1^2} + \lambda_2\frac{\left|c_{1,i} - c_{2,i}\right|}{\sigma_2^2}\right)\right] \\ &= \frac{1}{N}\sum_{i=1}^N \kappa^M_{c,\sigma}\left(c_{1,i} - c_{2,i}\right) = \frac{1}{N}\sum_{i=1}^N \kappa^M_{c,\sigma}(e_i).\end{aligned} \tag{4.5}$$

4.3 Property

Then, the based properties of the C–C are shown as follows:

Property 1 *The symmetry, i.e.,* $V^M_{c,\sigma}(c_1, c_2) = V^M_{c,\sigma}(c_2, c_1)$.

Property 2 $V^M_{c,\sigma}(c_1, c_2)$ *is positive and bounded, i.e.,* $0 < V^M_{c,\sigma}(c_1, c_2) \le 1$, *when and only when* $c_1 = c_2$, $V^M_{c,\sigma}$ *reaches the maximum.*

Property 3 *It is assumed that the samples* $\{c_{1,i}, c_{2,i}\}_{i=1}^N$ *obey joint PDF* $P_{C_1,C_2}(c_1, c_2)$, *and let* $\hat{P}_e(\cdot)$ *is the Parzen estimation of the error PDF with the samples* $\{e_i = c_{1,i} - c_{2,i}\}_{i=1}^N$, *and let* $V^M_{c,\sigma}$ *as Parzen window kernel.* $\hat{V}^M_{c,\sigma}(c_1, c_2)$ *is the value at zero point, i.e.,*

$$\hat{V}^M_{c,\sigma}(c_1, c_2) = \hat{P}_e(0). \tag{4.6}$$

Proof According to (4.5)

$$\begin{aligned}\hat{V}^M_{c,\sigma}(c_1, c_2) &= \frac{1}{N}\sum_{i=1}^{N} \kappa^M_{c,\sigma}\left(c_{1,i} - c_{2,i}\right) \\ &= \frac{1}{N}\sum_{i=1}^{N} \kappa^M_{c,\sigma}(e_i) \\ &= \hat{P}_e(0).\end{aligned} \tag{4.7}$$

where $\hat{P}_e(\varsigma) = \frac{1}{N}\sum\limits_{i=1}^{N} \kappa^M_{c,\sigma}(\varsigma - e_i)$.

Property 4 *When* $0 \le \lambda_1 \le 1$, $V^M_{c,\sigma}$ *is a second-order statistic on mapped feature space data.*

Proof When $0 \le \lambda_1 \le 1$, the kernel function $V^M_{c,\sigma}(c_1, c_2)$ still is a Mercer kernel, thus, $V^M_{c,\sigma}(c_1, c_2) = E[\varphi^M_{c,\sigma}(c_1)^H \varphi^M_{c,\sigma}(c_2)]$, where $\varphi^M_{c,\sigma}(\cdot)$ is a nonlinear map of $\kappa^M_{c,\sigma}$.

In the data analysis of regression as well as classification, the CMC-loss function in this chapter, inspired by a measure as GC-loss function [11], is defined as:

$$J_{CMC-loss}(c_1, c_2) = \kappa^M_{c,\sigma}(0) - V^M_{c,\sigma}(c_1, c_2). \tag{4.8}$$

Above CMC-loss satisfy $J_{CMC-loss}(c_1, c_2) \ge 0$, and it can be expressed as:

$$J_{CMC-loss}(c_1, c_2) = \frac{1}{2}E\left[\left\|\varphi^M_{c,\sigma}(c_1) - \varphi^M_{c,\sigma}(c_2)\right\|^2\right]. \tag{4.9}$$

Obviously, Minimizing CMC-loss is equivalent to maximizing the C–C. Thus, the CMC-loss is a meaningful mean square loss function on the feature space induced by the Mercer kernel $\kappa(c_1, c_2)$. Let $\{c_{1,i}, c_{2,i}\}_{i=1}^N$ is the N samples obtained from $P_{c_{1,i},c_{2,i}}(c_{1,i}, c_{2,i})$, the estimation of the CMC-loss is:

$$\begin{aligned}\hat{J}_{CMC-loss}\left(c_{1,i}, c_{2,i}\right) &= \kappa_{c,\sigma}^M(0) - \hat{V}_{c,\sigma}^M\left(c_{1,i}, c_{2,i}\right)\\ &= 1 - \frac{1}{N}\sum_{i=1}^N \kappa_{c,\sigma}^M\left(c_{1,i} - c_{2,i}\right)\\ &= 1 - \frac{1}{N}\sum_{i=1}^N \kappa_{c,\sigma}^M(e_i).\end{aligned} \tag{4.10}$$

Property 5 *Let* $\mathbf{X} = [x_1, x_2, ..., x_n]^T$, $\mathbf{Y} = [y_1, y_2, ..., y_n]^T$. *Then the combined correntropy induced metric (CCIM) is given by:* $\mathrm{CCIM}(\mathbf{X}, \mathbf{Y}) = \sqrt{\hat{J}_{CMC-loss}(\mathbf{X}, \mathbf{Y})}$ *that defines an n-dimensional sample vector space. where* $[\cdot]^T$ *denotes transpose operation.*

Proof Constructing a nonlinear mapping $\Phi_{c,\sigma}^M(\mathbf{X}) = [\varphi_{c,\sigma}^M(x_1)^T, ..., \varphi_{c,\sigma}^M(x_n)^T]^T$ and let $\mathbf{Z} = [z_1, z_2, ..., z_n]^T$. Thus, $||\Phi_{c,\sigma}^M(\mathbf{X}) - \Phi_{c,\sigma}^M(\mathbf{Y})|| = \sqrt{2N} \times \mathrm{CCIM}(\mathbf{X}, \mathbf{Y})$. There the CCIM is a metric because it satisfies three conditions: (i) The symmetry, i.e., $\mathrm{CCIM}(\mathbf{X}, \mathbf{Y}) = \mathrm{CCIM}(\mathbf{Y}, \mathbf{X})$; (ii) the triangular nature, i.e., $\mathrm{CCIM}(\mathbf{X}, \mathbf{Z}) \leq \mathrm{CCIM}(\mathbf{X}, \mathbf{Y}) + \mathrm{CCIM}(\mathbf{Y}, \mathbf{Z})$; (iii) The positive definiteness, i.e., $\mathrm{CCIM}(\mathbf{X}, \mathbf{Y}) \geq 0$ (The proof of these conditions is simple and will not be repeated here). Therefore, the CCIM function $\mathrm{CCIM}(\mathbf{X}, \mathbf{Y})$ defines a "Euclidean distance" in Hilbert space F_κ^N.

4.4 Adaptive Filtering Based on Combinatoric Correntropy Criterion

4.4.1 The Cost Function

Similar to MCC, the C-MCC can also be used as a cost function for estimating similarity. In the linear filtering, according to C-MCC, the optimal weight vector of the filter can be obtained by solving the minimized CMC-loss function; that is

$$\begin{aligned}J_{CMC-loss} &= \kappa_{c,\sigma}^M(0) - V_{c,\sigma}^M(e(m))\\ &= 1 - E\left\{\exp\left[-(\lambda_1\frac{e^2(m)}{2\sigma_1^2} + \lambda_2\frac{|e(m)|}{\sigma_2^2})\right]\right\},\end{aligned} \tag{4.11}$$

where $e(m) = d(m) - y(m) = d(m) - \boldsymbol{w}^T(m)\boldsymbol{x}(m)$ is the error vector, $d(m) \in \mathbb{R}$ denotes the desired value at time instant m, $y(m) = \boldsymbol{w}^T(m)\boldsymbol{x}(m)$ is the output, $\boldsymbol{w} = [w_1, w_2, ..., w_l]^T \in \mathbb{R}^{l\times 1}$ is the weight vector, $\boldsymbol{x}(m) = [x_1, x_2, ..., x_l]^T \in \mathbb{R}^{l\times 1}$ represents the input signal.

4.4.2 *The Adaptive Algorithm*

Recalling (4.11), the C-MCC algorithm according to the stochastic gradient descent can be formulated as follows:

$$\begin{aligned} w(m+1) &= w(m) + \mu \frac{\partial}{\partial w(m)} \left\{ 1 - \exp\left[-\left(\lambda_1 \frac{e^2(m)}{2\sigma_1^2} + \lambda_2 \frac{|e(m)|}{\sigma_2^2} \right) \right] \right\} \\ &= w(m) + \eta \exp\left[-\left(\lambda_1 \frac{e^2(m)}{2\sigma_1^2} + \lambda_2 \frac{|e(m)|}{\sigma_2^2} \right) \right] e(m)x(m), \end{aligned} \quad (4.12)$$

where $\eta = \mu(\frac{\lambda_1}{2\sigma_1^2} + \frac{\lambda_2}{\sigma_2^2 e(m)})$ is the learning rate.

4.5 Analysis of the C-MCC Algorithm

In this part, the theoretical analysis of the C-MCC algorithm will be derived in detail. The Eq. (4.12) can be rewritten as:

$$w(m+1) = w(m) + \eta g(e(m))x(m), \quad (4.13)$$

where $g(e(m)) = \exp[-(\lambda_1 \frac{e^2(m)}{2\sigma_1^2} + \lambda_2 \frac{|e(m)|}{\sigma_2^2})]e(m)$.

It is assumed that the desired signal is expressed as:

$$d(m) = w_0^T x(m) + v(m), \quad (4.14)$$

where w_0 is the estimated weight vector of the unknown system. Defining the weight error vector as: $\tilde{w}(m) = w_0 - w(m)$, $v(m)$ is the background noise. Thus, the error also can be expressed as:

$$\begin{aligned} e(m) &= \tilde{w}^T(m-1)x(m) + v(m) \\ &= e_a(m) + v(m). \end{aligned} \quad (4.15)$$

where $e_a(m)$ is defined as a *priori error*. Combining (4.12), (4.15) and $\tilde{w}(m) = w_0 - w(m)$, we have

$$\begin{aligned} E\left[\|\tilde{w}(m+1)\|^2\right] &= E\left[\|\tilde{w}(m)\|^2\right] - 2\mu E\left[g(e(m))e_a(m)\right] \\ &+ \mu^2 E\left[g^2(e(m))\|x(m)\|^2\right]. \end{aligned} \quad (4.16)$$

To facilitate the analysis, the following assumptions are needed:

A: $v(m)$ is zero-mean with the variance is σ_v^2, independent, identically, distributed, and is independent of the input $x(m)$ [17–19].

B: $e_a(m)$ is zero-mean and independent of the noise $v(m)$ [19].

C: the filter is long enough, in this condition, $e_a(m)$ is Gaussian and $||\boldsymbol{x}(m)||^2$ is asymptotically independent of $g^2(e(m))$ [20].

The Eq. (4.17) can be derived via employing the above assumptions, that is

$$\lim_{i\to\infty} E\big[\big|g^2(e(m))\big|\|\boldsymbol{x}(m)\|^2\big] = \mathrm{Tr}(\boldsymbol{R}_{\mathbf{XX}})\lim_{i\to\infty} E\big[\big|g^2(e(m))\big|\big], \tag{4.17}$$

where $\boldsymbol{R}_{\mathbf{XX}} = E[\boldsymbol{x}(m)\boldsymbol{x}^T(m)]$, and then assuming the filter is stable and reaches the steady state. There have:

$$\lim_{i\to\infty} E\big[\|\tilde{\boldsymbol{w}}(m+1)\|^2\big] = \lim_{i\to\infty} E\big[\|\tilde{\boldsymbol{w}}(m)\|^2\big]. \tag{4.18}$$

And according to (4.16):

$$2\lim_{i\to\infty} E\big[g(e(m))e_a(m)\big] = \mathrm{Tr}(\boldsymbol{R}_{\mathbf{XX}})\eta\lim_{i\to\infty} E\big[g^2(e(m))\big]. \tag{4.19}$$

Defining the expectation of the mean-square of a priori *error*, i.e., $E[e_a^2(m)]$ is the steady-state excess mean square error (EMSE). In the next part, using the T_S denotes the theoretical value, i.e., $T_S = \lim_{i\to\infty} E[e_a^2(m)]$ that is obtained by Eq. (4.16).

By employing the assumptions A, B, C and taking the Taylor expansion of $g(e(m))$ with respect to e_a around v, that is

$$\begin{aligned} g(e(m)) &= g(e_a + v) \\ &= g(v) + g'(v)e_a + \frac{1}{2}g''(v)e_a^2 + o\big(e_a^2\big), \end{aligned} \tag{4.20}$$

where $g'(v)$ and $g''(v)$ are the first and second derivatives, and $o(e_a^2)$ is the third and higher-order terms. When the expectation of $o\big(e_a^2\big)$ is enough small in (4.20), and using the assumptions of A and B, there have

$$E\big[g(e)e_a\big] = E\big[g(v)e_a + g'(v)e_a^2 + o\big(e_a^2\big)\big] \approx E\big[g'(v)\big]T_S, \tag{4.21}$$

$$E\big[g^2(e)\big] \approx E\big[g^2(v)\big] + E\Big[g(v)g''(e) + \big|g'(v)\big|^2\Big]T_S. \tag{4.22}$$

Substitute (4.21), (4.22) into (4.20) yields

$$\begin{aligned} 2\mathrm{E}\big[g'(v)\big]T_S &= \eta\mathrm{Tr}(\boldsymbol{R}_{\mathbf{XX}})\Big\{E\big[g^2(v)\big] + E\Big[g(v)g''(e) + \big|g'(v)\big|^2\Big]T\Big\}_S \\ \Rightarrow T_S &= \frac{\eta\mathrm{Tr}(\boldsymbol{R}_{\mathbf{XX}})E\big[g^2(v)\big]}{2\mathrm{E}\big[g'(v)\big] - \eta\mathrm{Tr}(\boldsymbol{R}_{\mathbf{XX}})E\big[g(v)g''(e) + |g'(v)|^2\big]} \\ &= \frac{\eta\mathrm{Tr}(\boldsymbol{R}_{\mathbf{XX}})E\Big\{\exp\Big[-2\Big(\lambda_1\frac{v^2}{2\sigma_1^2} + \lambda_2\frac{|v|}{\sigma_2^2}\Big)\Big]v^2\Big\}}{\psi(v) - \eta\mathrm{Tr}(\boldsymbol{R}_{\mathbf{XX}})E[\varphi(v)]}, \end{aligned} \tag{4.23}$$

where

$$\psi(v) = 2\mathrm{E}\left\{\exp\left[-\left(\lambda_1 \frac{v^2}{2\sigma_1^2} + \lambda_2 \frac{|v|}{\sigma_2^2}\right)\right](1-v)\left(\frac{\lambda_1 v}{\sigma_1^2} + \frac{\lambda_2}{\sigma_2^2}\right)\right\}, \tag{4.24}$$

$$\begin{aligned}\varphi(v) = &\left|\exp\left[-\left(\lambda_1 \frac{v^2}{2\sigma_1^2} + \lambda_2 \frac{|v|}{\sigma_2^2}\right)\right](1-v)\left(\frac{\lambda_1 v}{\sigma_1^2} + \frac{\lambda_2}{\sigma_2^2}\right)\right|^2 \\ &+ \exp\left[-2\left(\lambda_1 \frac{v^2}{2\sigma_1^2} + \lambda_2 \frac{|v|}{\sigma_2^2}\right)\right]\zeta(v),\end{aligned} \tag{4.25}$$

where

$$\zeta(v) = \frac{\lambda_1^2 v^4}{\sigma_1^4} + \frac{2\lambda_1\lambda_2 v^3}{\sigma_1^2\sigma_2^2} + \frac{\lambda_2^2 v^2}{\sigma_2^2} - \frac{3\lambda_1 v^2}{\sigma_1^2} - \frac{2\lambda_2 v}{\sigma_2^2}. \tag{4.26}$$

When the learning step size is small enough, it leads to

$$T_S \approx \frac{\eta \mathrm{Tr}(\boldsymbol{R}_{\mathbf{XX}}) E\left\{\exp\left[-2\left(\lambda_1 \frac{v^2}{2\sigma_1^2} + \lambda_2 \frac{|v|}{\sigma_2^2}\right)\right] v^2\right\}}{2E\left\{\exp\left[-\left(\lambda_1 \frac{v^2}{2\sigma_1^2} + \lambda_2 \frac{|v|}{\sigma_2^2}\right)\right](1-v)\left(\frac{\lambda_1 v}{\sigma_1^2} + \frac{\lambda_2}{\sigma_2^2}\right)\right\}}. \tag{4.27}$$

4.6 Computer Simulation Examples

In this part, the performance of the C-MCC algorithm is compared with MCC, GMCC, and other algorithms, and the theoretical analysis of the proposed algorithm is verified. $\boldsymbol{w}_0 = [0.1, 0.21, 0.32, 0.43, 0.54, 0.43, 0.32, 0.21, 0.1]^T$ is the system weight vector to be estimated. Considering the practical situation, the noise is generated in three ways: one is generated by the noise model I, i.e., $v(m) = \alpha(m)Z_1(m) + Z_2(m)$. And one is generated by the model II, i.e., $v(m) = (1-\beta(m))Z_3(m) + \beta(m)Z_4(m)$, where $\alpha(m)$ is a Bernoulli–Gaussian process, $\beta(m)$ is a binary independent and identically distributed process with $\Pr\{\alpha(m) = 1\} = p_1$, $\Pr\{\alpha(m) = 0\} = 1 - p_1$, $\Pr\{\beta(m) = 1\} = p_2$, $\Pr\{\beta(m) = 0\} = 1 - p_2$, and $0 \le p_1, p_2 \le 1$ are the probability of occurrence; $Z_1(m)$, $Z_2(m)$, $Z_4(m)$ are white Gaussian noises with zero-mean and variance 10, 0.01, and 20 (very large variances represent large outliers), respectively, $Z_3(m)$ obeys Gaussian distribution with zero-mean and variance 0.1. The noise processes $Z_1(m)$ and $Z_2(m)$ are mutually independent and they are both independent of $\alpha(m)$; $Z_3(m)$ and $Z_4(m)$ are mutually independent and they are both independent of $\beta(m)$. And other is generated by the model III that the noise is random noise, i.e., it is randomly generated by Uniform distribution over the interval [-0.5, 0.5]. The MSD is used to test the behaviors of the

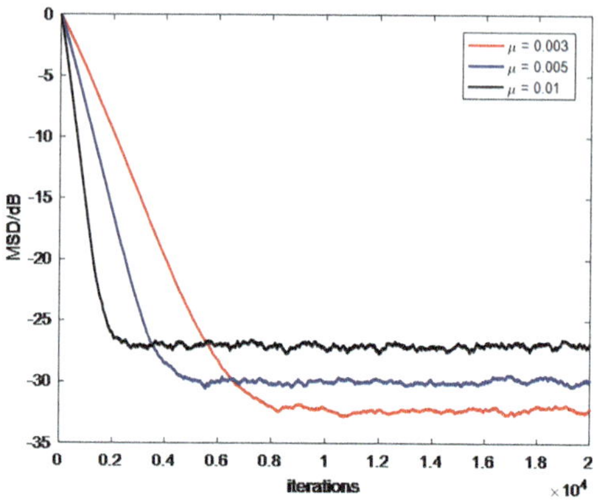

Fig. 4.1 Learning curves with different step sizes μ

algorithms, where MSD denotes as: $\text{MSD} = 10\log_{10}(\|\boldsymbol{w}_0 - \boldsymbol{w}\|_2^2 / \|\boldsymbol{w}\|_2^2)$, EMSE is employed to verify the correctness of the theoretical analysis method.

4.6.1 Effect on the C-MCC Algorithm of Different Parameters

The behaviors of the C-MCC algorithm with different step sizes, kernel widths, and the mixture parameters are compared in this section, the results are shown in Figs. 4.1, 4.2 and 4.3. It can be seen that the convergence speed of the proposed algorithm is proportional to the increase or decrease of the step size, and the steady-state performance is inversely proportional to the change of the step size; as the kernel width increases simultaneously, the convergence speed becomes slower, but steady-state performance gradually improves; for the change of the combined parameters, the performance of the algorithm is better when λ_1 occupies a heavier proportion.

4.6.2 The Adaptive Algorithm

In this section, we choose $p_1 = 0.1$, $p_2 = 0.07$ to generate the noise for testing the performance of the algorithms. Figure 4.4 shows the performance comparison between the C-MCC algorithm and MCC, GMCC algorithms, and the results of tracking behaviors of the C-MCC, MMCC and other based MCC algorithms are shown in Figs. 4.5, 4.6 and 4.7, where the noise interference in Fig. 4.5 is generated by the noise model I, and that in Fig. 4.6 is generated by the noise model II, the noise interference is generated by the noise model III in Fig. 4.7. It can be seen from Fig. 4.4 that the C-MCC algorithm has better steady-state performance than MCC, GMCC algorithms. The tracking performance of the proposed algorithm is

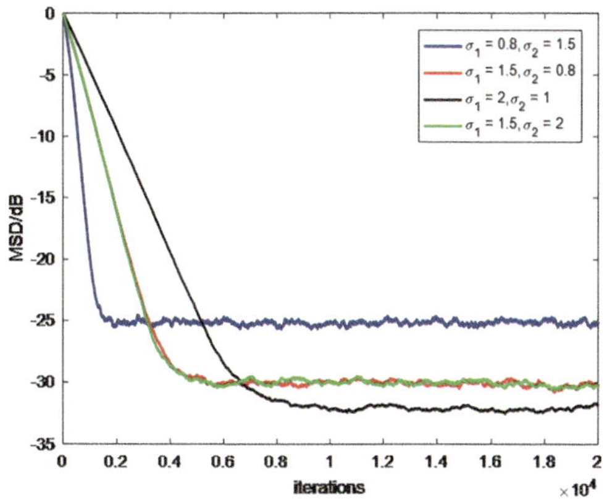

Fig. 4.2 Learning curve with different kernel widths σ

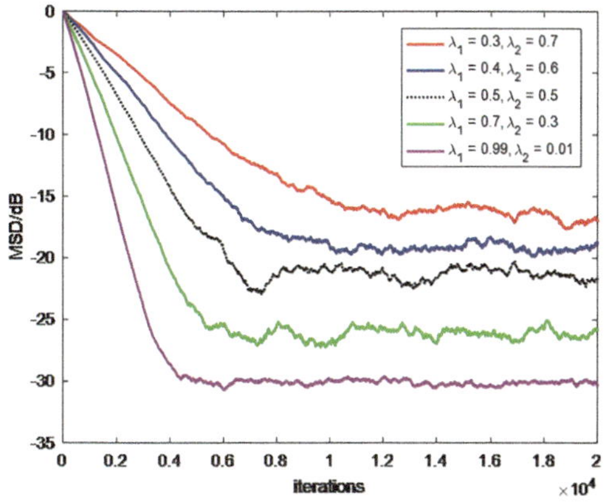

Fig. 4.3 Learning curves with different combined parameters λ_1,λ_2

also compared in experiments, it can be seen that the C-MCC algorithm can quickly track sudden changes in the system, the convergence speed is faster than the based-MCC algorithms and has lower the steady-state MSD than other algorithms under three different noise interference simulation conditions.

Then the steady-state MSD of the C-MCC and other different algorithms is summarized in Table 4.1. It can be seen from Table 4.1 that the proposed algorithm has the lowest steady-state MSD than other algorithms.

4.6.3 Theoretical Verification

In this section, the theoretical analysis of the proposed algorithm is verified. Figure 4.8 shows the comparison plot under different choices of step values. It can be seen that the value of EMSE increases as the step size increases, and the theoretical and the

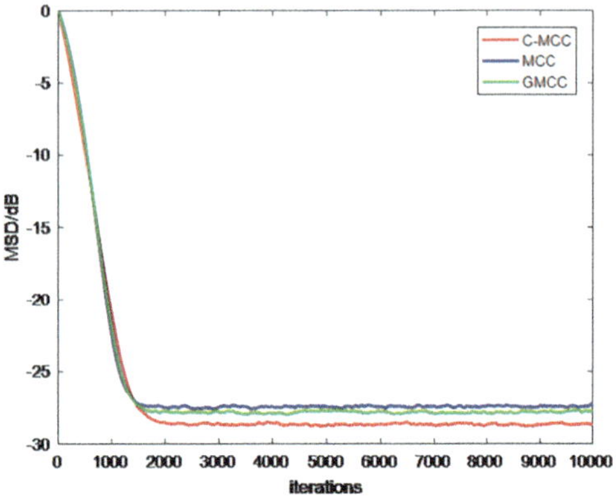

Fig. 4.4 Comparison curves between the C-MCC and other MCC algorithms

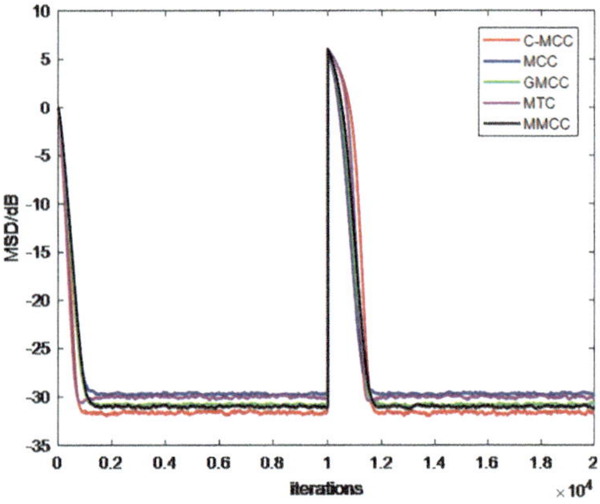

Fig. 4.5 Tracking curves between the C-MCC and other algorithms with noise interference generated by noise model I

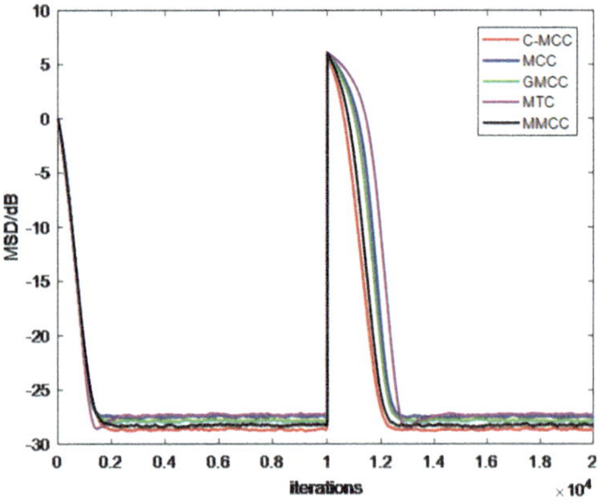

Fig. 4.6 Tracking curves between the C-MCC and other algorithms with the noise interference generated by the noise model II

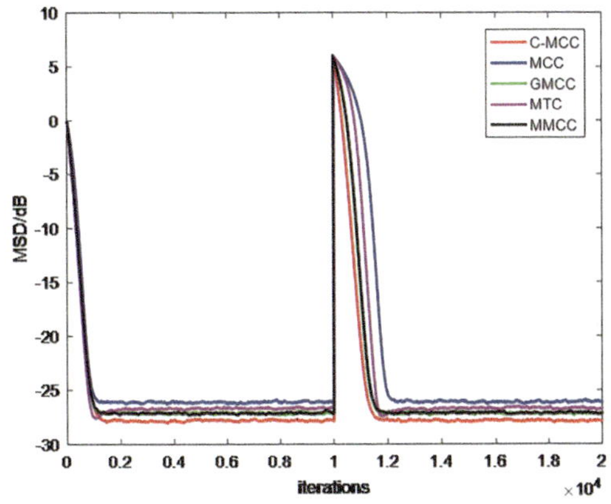

Fig. 4.7 Tracking curves between the C-MCC and other algorithms with the noise interference generated by the noise model III

Table 4.1 Steady-state msd of the different algorithms

Algorithms	Noise model I	Noise model II	Noise model III
MCC	0.0011	0.0019	0.0025
MMCC	0.0008	0.0015	0.0019
GMCC	0.0009	0.0017	0.0020
MTC	0.0010	0.0014	0.0022
C-MCC	0.0007	0.0019	0.0017

experimental value are in high agreement, which shows the correctness and validity of the theoretical analysis.

Figure 4.9 shows a comparison of the experimental simulation of the theoretical analysis under different noise variances. It can be seen that the value of EMSE is proportional to the variation of noise variance, and the experimental value can match well with the theoretical curve.

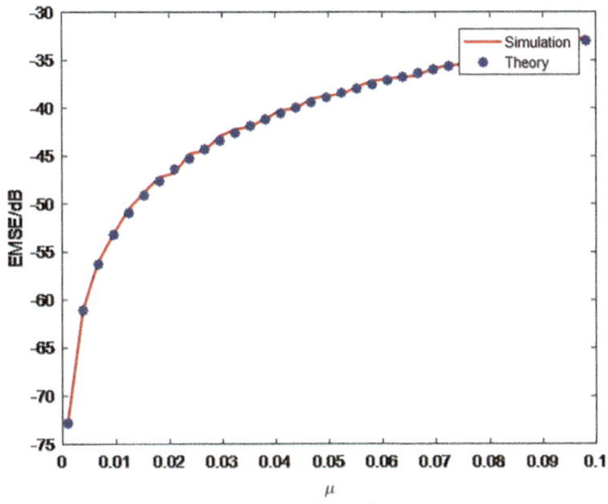

Fig. 4.8 Comparison curves of theory and simulation with different step sizes

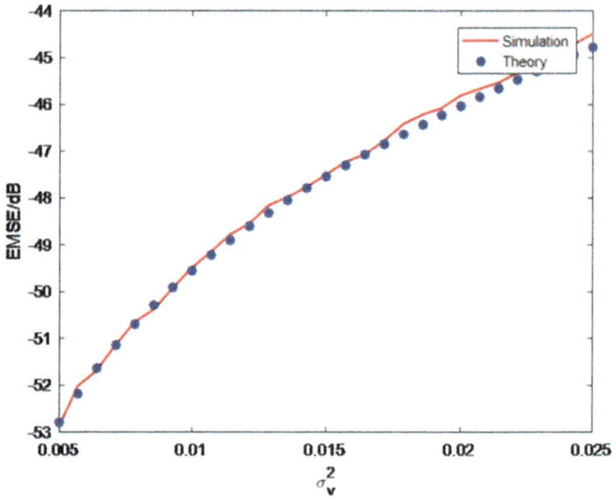

Fig. 4.9 Comparison curves of theory and simulation with different variances

4.7 Summary

In this chapter, a new form of correntropy based on combining two single kernels called the C-MCC is given and its adaptive filtering algorithm is proposed. The basic properties of the C-C have been elaborated by the theoretical analysis, and the influence of the proportion of the combined parameters on the algorithm performance is also discussed. Moreover, the steady-state theoretical performance of the C-MCC algorithm is analyzed under some simple assumptions and by employing the Taylor expansion method. Finally, the superiority of the proposed algorithm and the accuracy of the theoretical analysis method are demonstrated in the simulation section.

References

1. S. Haykin, *Adaptive Filter Theory*, 5th edn. (Pearson, London, U.K., 2014)
2. H. Sayed, *Fundamentals of Adaptive Filtering* (Wiley, Hoboken, NJ, USA, 2003)
3. J. Benesty, H. Rey, L.R. Vega, S. Tressens, A nonparametric VSS NLMS algorithm. IEEE Signal Process. Lett. **13**(10), 581–584 (2006)
4. N.J. Bershad, J.C.M. Bermudez, J. Tourneret, An affine combination of two LMS adaptive filters—transient mean-square analysis. IEEE Trans. Signal Process. **56**(5), 1854–1864 (2008)
5. G. Wang, H. Zhao, Robust adaptive least mean M-estimate algorithm for censored regression. IEEE Trans. Syst. Man Cybern. Syst. **52**(8), 5166–5174 (2022)
6. W. Liu, P.P. Pokharel, J.C. Principe, Correntropy: properties and applications in non-Gaussian signal processing. IEEE Trans. Signal Process. **55**(11), 5286–5298 (2007)
7. Singh, J.C. Principe, Using correntropy as a cost function in linear adaptive filters. Proc. IEEE Int. Joint Conf. Neural Netw. (IJCNN), pp. 2950–2955 (2009)
8. Chen, X. Wang, Y. Li, J.C. Principe, Maximum correntropy criterion with variable center. IEEE Signal Process. Lett. **26**(8), 1212–1216 (2019)
9. R. He, B.-G. Hu, W.-S. Zheng, X.-W. Kong, Robust principal component analysis based on maximum correntropy criterion. IEEE Trans. Image Process. **20**(6), 1486.1494 (2011)
10. Erdogmus, J.C. Principe, Generalized information potential criterion for adaptive system training. IEEE Trans. Neural Netw. **13**(5), 1036–1044 (2002)

11. Chen, L. Xing, H. Zhao, N. Zheng, J.C. Principe, Generalized Correntropy for Robust Adaptive Filtering. IEEE Trans. Signal Process. **64**(13), 3377–3387 (2016)
12. F. Wang, Y. He, S. Wang, B. Chen, Maximum total correntropy adaptive filtering against heavy-tailed noises. Signal Process. **141**, 84–95 (2017)
13. Chen, X. Wang, N. Lu, S. Wang, J. Cao, J. Qin, Mixture correntropy for robust learning. Pattern Recognit. **79**, 318–327 (2018)
14. H. Zhao, B. Liu, P. Song, Variable step-size affine projection maximum correntropy criterion adaptive filter with correntropy induced metric for sparse system identification. IEEE Trans. Circuits Syst. II, Exp. Briefs **67**(11), 2782–2786 (2020)
15. P. Song, H. Zhao, Affine-projection-like M-estimate adaptive filter for robust filtering in impulse noise. IEEE Trans. Circuits Syst. II, Exp. Briefs **66**(12), 2087–2091 (2019)
16. Y. Zhu, H. Zhao, X. Zeng, B. Chen, Robust generalized maximum correntropy criterion algorithms for active noise control. IEEE/ACM Trans. Audio, Speech, Lang. Process. **28**, 1282–1292 (2020)
17. W. Shi, Y. Li, Y. Wang, Noise-free maximum correntropy criterion algorithm in non-gaussian environment. IEEE Trans. Circuits Syst. II, Exp. Briefs **67**(10), 2224–2228 (2020)
18. H. Zhao, D. Liu, S. Lv, Robust maximum correntropy criterion subband adaptive filter algorithm for impulsive noise and noisy input. IEEE Trans. Circuits Syst. II, Exp. Briefs **69**(2), 604–608 (2022)
19. W. Wang, H. Zhao, X. Zeng, K. Doğançay, Steady-state performance analysis of nonlinear spline adaptive filter under maximum correntropy criterion. IEEE Trans. Circuits Syst. II, Exp. Briefs **67**(6), 1154–1158 (2020)
20. B. Chen, L. Xing, J. Liang, N. Zheng, J.C. Príncipe, Steady-state mean-square error analysis for adaptive filtering under the maximum correntropy criterion. IEEE Signal Process. Lett. **21**(7), 880–884 (2014)

Chapter 5
Fast Minimum Error Entropy with Fiducial Points

Recently, there has been a strong interest in the minimum error entropy (MEE) criterion derived from information theoretic learning, which is effective in dealing with the multimodal non-Gaussian noise case. However, the kernel function is shift invariant resulting in the MEE criterion being insensitive to the error location. An existing solution is to combine the maximum correntropy (MC) criterion with MEE, leading to the minimum error entropy criterion with fiducial points (MEEF). Nevertheless, the MEEF-based algorithms usually require higher computational complexity. To remedy this problem, a Fast MEEF (FMEEF) criterion is devised, aiming to avoid repetitive calculations of the a posteriori error, and an adaptive filtering algorithm based on gradient descent (GD) is proposed, namely, GD-based FMEEF (FMEEF-GD) algorithm. In addition, we provide the convergence condition in terms of mean sense, along with an analysis of the steady-state and transient behaviors of FMEEF-GD in the mean square sense. Its computational complexity analysis is also analyzed. Simulation results demonstrate that the computational requirement of our algorithm does not vary significantly with the error sample number and the derived theoretical model is highly consistent with the learning curve. Ultimately, we employ the FMEEF-GD algorithm in tasks such as system identification, wind signal magnitude prediction, and acoustic echo cancellation (AEC) to validate the effectiveness of the FMEEF-GD method.

5.1 Introduction

The performance of adaptive filtering algorithms relying on the minimum-mean-square-error (MMSE) criterion may degrade significantly in environments with non-Gaussian noise. Moreover, the existence of ubiquitous non-Gaussian noise has attracted considerable interest in studying the non-MMSE criterion to improve the algorithm robustness. Recently, the information theoretic measure (ITM) method,

H. Zhao et al., *Information Theoretic Learning-Based Filter*,
https://doi.org/10.1007/978-3-032-29623-8_5

with entropy as a metric, has found widespread application in machine learning and signal processing, specifically in information theoretic learning (ITL). Both minimum error entropy (MEE) and maximum correntropy (MC) algorithms have been developed based on ITL and have excellent performance in impulsive noise environments [1–6]. A primary reason why ITL has gained widespread attention is that ITM can capture more statistical information about the data and provide potential performance improvements in machine learning applications. For noise with more complicated distributions, such as a multimodal distribution, the MEE algorithm has better performance than the MC algorithm [7], since the former criterion can accurately fit the error distribution by utilizing the Parzen window method.

However, the widely used cost function of MEE involves maximizing the quadratic information potential (QIP) leading to high computational complexity [4]. To tackle this problem, the quantized MEE (QMEE) criterion was proposed, whose core concept is to streamline the inner summation process by quantizing the error samples for complexity reduction [8]. In fact, the computational complexity of the QMEE criterion remains significantly influenced by the number of codebooks. In addition, the shift invariance of the kernel function causes the MEE criterion to be insensitive to the error's mean value and makes it difficult to obtain an optimal solution. The MEE criterion does not yield a closed-form solution even for a straightforward linear system. Different approaches have been suggested to seek for the optimal solution. For example, the results of MEE are corrected by setting a bias to produce a zero mean error after training. Another method is to combine the location-sensitive MC into MEE, resulting in a minimum error entropy criterion with fiducial points (MEEF) that automatically adjusts the error mean to the origin [9–12]. Compared with the MEE criterion, MEEF tends to achieve better performance.

Recently, many algorithms based on MEEF have been developed [10, 13]. However, since MEEF-based algorithms usually exhibit high computational complexity [14, 15], this limits their capabilities in scenarios where real-time processing is required. To solve this problem, a Fast MEEF (FMEEF) criterion is proposed in this paper. This criterion only considers the relationship between the error at time m and the mean of N error samples within the window, without taking into account the relationships among all the errors within the window. As a result, the computational complexity of the criterion is significantly reduced. Unlike the QMEE criterion, our method is almost independent of the number of samples. In other words, the FMEEF has a lower computational complexity. Based on this criterion, we develop an FMEEF algorithm based on gradient descent method (FMEEF-GD). In addition, the FMEEF-GD algorithm requires frequent computation of the a posteriori error at each iteration. To alleviate this computational burden, we utilize the a priori error to extract the error mean information during the iteration process, which further reduces the computational complexity. Finally, we analyze the convergence condition, transient and steady-state behaviors, and complexity of the FMEEF-GD algorithm. The primary contributions of our work are summarized as follows.

(1) To address the computational complexity issues associated with the MEEF criterion [15], we present the FMEEF criterion and subsequently develop an FMEEF-GD algorithm based on this criterion.
(2) The convergence condition, transient performance, steady-state performance of the FMEEF-GD algorithm are analyzed based on the moment generating function (MGF) [16], and the computational complexity analysis is provided [17].
(3) The simulation results confirm that the FMEEF-GD algorithm has excellent performance in system identification, wind signal prediction, and acoustic echo cancellation (AEC) compared to the least mean square (LMS) [17], MC [16], and MEEF-GD [11] algorithms.

5.2 Minimum Error Entropy with Fiducial Points

In this section, we review the MC, MEE and MEEF criteria. Subsequently, we derive the MEEF-GD algorithm based on the MEEF criterion.

5.2.1 Maximum Correntropy Criterion

Correntropy serves as a similarity measure between two random variables D and Y [2, 6], which is defined as:

$$V(D, Y) = \mathrm{E}[k_\sigma(D, Y)] \tag{5.1}$$

where $\mathrm{E}[\cdot]$ denotes the expectation operator, $k_\sigma(\cdot, \cdot)$ is a kernel function with bandwidth σ. A popular choice for the kernel function is the Gaussian kernel, which is defined as:

$$k_\sigma(D, Y) = \exp\left(-\frac{(D-Y)^2}{2\sigma^2}\right) \tag{5.2}$$

In practice, correntropy is commonly estimated using a limited number of available data samples due to the unknown joint distribution of D and Y. The correntropy sample estimator based on N samples $\{d(i), y(i)\}_{i=1}^N$ of random variables D and Y is

$$\hat{V}(D, Y) = \frac{1}{N}\sum_{i=1}^{N} k_\sigma(d(i), y(i)) = \frac{1}{N}\sum_{i=1}^{N} k_\sigma(e(i)) \tag{5.3}$$

where N represents the sample number, and $e(i) = d(i) - y(i)$.. Consequently, the MC criterion obtains the optimal solution by maximizing (5.3).

5.2.2 Minimum Error Entropy Criterion

Since entropy reflects the degree of uncertainty in a random variable, the MEE criterion actually makes the prediction error more concentrated [18, 19]. The second-order Renyi's entropy is usually employed in MEE to quantify the error information, which is expressed as:

$$H(e) = -\log \int p^2(e)\mathrm{d}e = -\log(V(e)) \tag{5.4}$$

where E denotes the error between random variables D and Y, $p(e)$ represents the probability density function of E, and $V(e) = \int p^2(e)\mathrm{d}e$. Using the Parzen windows method to estimate $p(e)$, the entropy sample estimator based on N samples is computed as:

$$\hat{V}(e) = \int \hat{p}_E^2(e)\mathrm{d}e = \frac{1}{N^2}\sum_{i=1}^{N}\sum_{j=1}^{N} k_{\sigma\sqrt{2}}(e(i) - e(j)) \tag{5.5}$$

One can observe that MEE criterion is actually equivalent to maximizing (5.5). Due to the shift-invariant property of the MEE criterion, it becomes challenging to determine the location of the error mean. As a result, it is essential to introduce a bias into the MEE criterion to avoid this problem.

5.2.3 Minimum Error Entropy with Fiducial Points

The MEEF criterion was developed to enhance the sensitivity of the MEE criterion to the error mean by integrating the MC with the MEE in [9]:

$$J(e) = \lambda \sum_{i=1}^{N} k_{\sigma_1}(e(i)) + (1 - \lambda) \sum_{i=1}^{N}\sum_{j=1}^{N} k_{\sigma_2}(e(i) - e(j)) \tag{5.6}$$

where $\lambda \in [0, 1]$ indicates the weighting coefficient which determines the number of fiducial points placed [9, 10]. It is noteworthy that when $\lambda = 1$, MEEF reduces to MC, and when $\lambda = 0$, it reduces to MEE. In practice, the MC criterion drives the error peak close to zero, while the MEE criterion tends to concentrate the error distribution. Consequently, MEEF is typically superior to MC and MEE.

To illustrate how the optimal solution can be obtained from the above empirical risk function, we consider a typical linear system model:

$$d(m) = \boldsymbol{w}_{\mathrm{opt}}^{\mathrm{T}}\boldsymbol{x}(m) + v(m) \tag{5.7}$$

where $d(m)$ denotes the desired signal, $(\cdot)^{\mathrm{T}}$ refers to the transpose operator, $v(m)$ represents the additional noise, $\mathbf{w}_{\mathrm{opt}}$ is the L-dimension optimal weight vector, and $\boldsymbol{x}(m)$ stands for the input vector. Combined with GD, the weight vector of the adaptive filtering algorithm is updated as:

$$\boldsymbol{w}(m+1) = \boldsymbol{w}(m) + \mu \frac{\partial J(e)}{\partial \boldsymbol{w}} \tag{5.8}$$

where $\mu > 0$ is the step size and $\boldsymbol{w}(m)$ is the weight vector at time m. In online applications [20–22], the estimation of information potential can be obtained using the stochastic information gradient. Thus, the stochastic gradient of (5.8) is approximated by $J_N(e(m)) \approx J(e)$, and $J_N(e(m))$ is formulated as:

$$J_N(e(m)) = \lambda k_{\sigma_1}(e(m)) + (1-\lambda) \sum_{i=m-N}^{m-1} k_{\sigma_2}(e(m) - \in(m,i)) \tag{5.9}$$

where $e(m) = d(m) - \boldsymbol{w}(m)^{\mathrm{T}}\boldsymbol{x}(m)$ and $\in(m,i) = d(i) - \boldsymbol{w}(m)^{\mathrm{T}}\boldsymbol{x}(i)$. This approximation assumes that the stochastic process $e(m)$ is stationary. Therefore, we can derive the weight update equation of the MEEF-GD algorithm:

$$\begin{aligned}\boldsymbol{w}(m+1) = \boldsymbol{w}(m) &+ \frac{\mu\lambda}{\sigma_1^2} k_{\sigma_1}(e(m))e(m)\boldsymbol{x}(m) \\ &+ \frac{\mu(1-\lambda)}{\sigma_2^2} \sum_{i=m-N}^{m-1} k_{\sigma_2}(e(m) - \in(m,i))(e(m) - \in(m,i))(\boldsymbol{x}(m) - \boldsymbol{x}(i))\end{aligned} \tag{5.10}$$

Remark 5.1 As can be seen, the above update equation is the combination of the MC and MEE algorithms. It has been verified that lower steady-state errors and smaller error dispersion can be obtained by adjusting the weighting parameter λ compared to the MC and MEE algorithms [10]. However, it requires repeated calculation of error $\in(m,i)$ using the filter parameter vector $\boldsymbol{w}(m)$, which results in an increase in the algorithm complexity.

5.3 Fast Minimum Error Entropy with Fiducial Points

The computational drawback of MEEF-GD algorithm (5.10) makes it unsuitable for real-time applications. Therefore, we propose a Fast MEEF (FMEEF) criterion in this section, which can significantly reduce the computational complexity of the MEEF criterion. Moreover, the FMEEF-GD algorithm is developed.

The primary factor contributing to the high computational complexity of the MEEF criterion is the operation of the MEE part. To simplify the double summation process in the MEE part and achieve a concentration of the error distribution similar to the MEE criterion, we maintain the errors in close proximity to the mean of the errors. Therefore, the FMEEF criterion is expressed as

$$J(e) = \lambda \sum_{i=1}^{N} k_{\sigma_1}(e(i)) + (1 - \lambda) \sum_{i=1}^{N} k_{\sigma_2}(e(i) - \overline{e}(i)) \tag{5.11}$$

where $\overline{e}(i)$ denotes the error mean within sample window, and

$$\overline{e}(i) = \frac{1}{N} \sum_{j=i-N}^{i-1} \in (i, j) \tag{5.12}$$

In order to fit the online training algorithm and avoid the repetitive computation of error $\in (i, j)$, we use $e(j)$ instead of $\in (i, j)$, i.e., we use the a priori error instead of the a posteriori error. Since $e(j)$ is already obtained during the iterative process, this approach allows us to streamline the calculation process and utilize the existing error estimates for the error mean calculation, contributing to a more efficient and effective algorithm implementation. Therefore, $\overline{e}(i)$ is represented as

$$\overline{e}(i) = \frac{1}{N} \sum_{j=i-N}^{i-1} e(j) \tag{5.13}$$

To visualize the relationship between MEEF and FMEEF criteria, a two-tap finite impulse respond (FIR) filter with weight vector $\boldsymbol{w}_{\text{opt}} = [w_1, w_2]^{\text{T}} = [0, 0]^{\text{T}}$ is considered [23]. The performance surfaces and the gradient differences (grad[J_{MEEF}] − grad[J_{FMEEF}]) of the FMEEF and MEEF criteria with respect to w_1 and w_2 in a Gaussian white noise environment with variance 0.1 are shown in Figs. 5.1 and 5.2, where the parameters of both the MEEF and FMEEF criteria are set to $\sigma_1 = 1$, $\sigma_2 = 2$ and $\lambda = 0.5$. It can be seen from Fig. 5.1 that both MEEF and FMEEF criteria have a small gradient when far away from the optimal weights $\boldsymbol{w}_{\text{opt}}$, the gradient becomes larger when close to the optimal weights, and the gradient is close to 0 when the optimal weights are reached. MEEF and FMEEF behave very similarly from this perspective. Moreover, the MEEF criterion is superior to the FMEEF criterion in terms of the magnitude of the gradient in this case.

Figure 5.3 depicts the block diagram of adaptive filtering under the FMEEF criterion. To obtain the optimal solution under the FMEEF criterion, the GD method is adopted. According to the GD method and Fig. 5.3, the weights of the FMEEF filter are updated as

$$\boldsymbol{w}(m+1) = \boldsymbol{w}(m) + \xi_1 g_{\sigma_1}(e(m)) + \xi_2 g_{\sigma_2}(e(m) - \overline{e}(m)) \tag{5.14}$$

where $\xi_1 = \mu\lambda/\sigma_1^2$, $\xi_2 = \mu(1 - \lambda)/\sigma_2^2$,

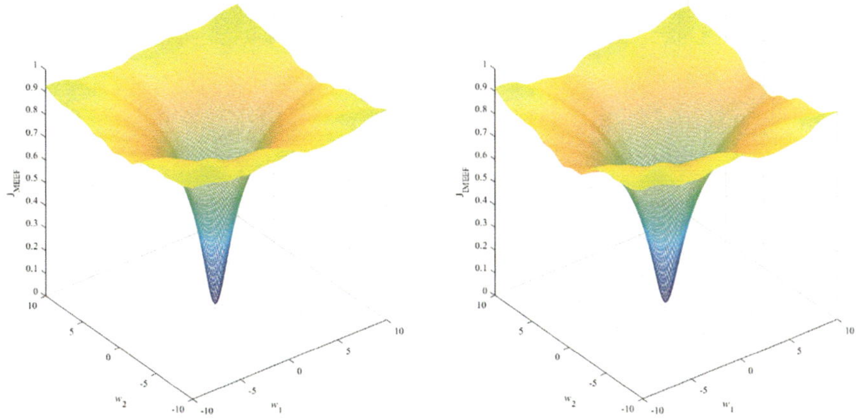

Fig. 5.1 Performance surfaces of MEEF (left) and FMEEF criteria (right)

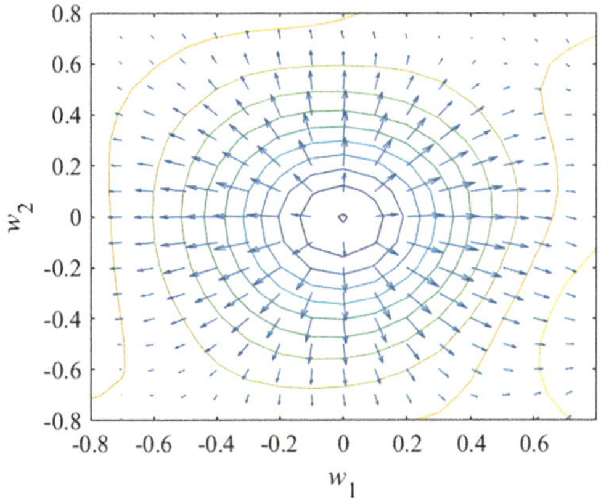

Fig. 5.2 Contour and gradient difference between MEEF and FMEEF criteria

$$g_{\sigma_1}(e(m)) = k_{\sigma_1}(e(m))e(m)\boldsymbol{x}(m) \tag{5.15}$$

$$g_{\sigma_2}(e(m) - \overline{e}(m)) = k_{\sigma_2}(e(m) - \overline{e}(m))(e(m) - \overline{e}(m))(\boldsymbol{x}(m) - \overline{\boldsymbol{x}}(m)) \tag{5.16}$$

where

$$\overline{\boldsymbol{x}}(m) = \frac{1}{N} \sum_{i=m-N}^{m-1} \boldsymbol{x}(i) \tag{5.17}$$

Remark 5.2 As one can observe from the iterative update Eq. (5.14), the proposed algorithm no longer needs to repeatedly compute a posteriori errors. It is also worth noting that the complexity of the FMEEF-GD algorithm is almost independent of the size of Parzen window N.

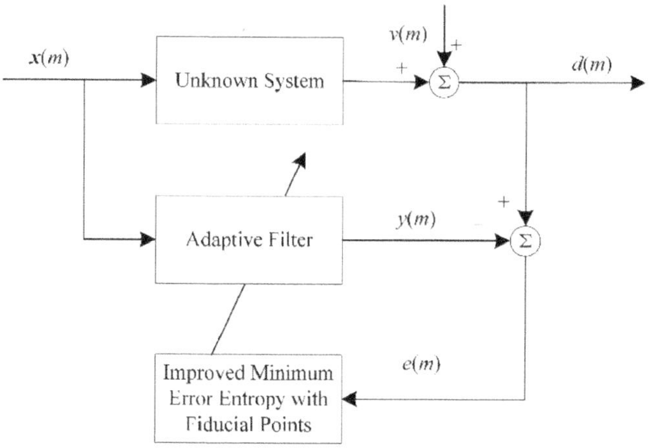

Fig. 5.3 Adaptive filtering under FMEEF criterion

To ensure that the error mean is obtained, only the MC portion is used in (5.14) at the first N iterations. Utilizing the previous derivation, the pseudocode for the FMEEF-GD algorithm is described in Algorithm 5.1.

Remark 5.3 When the problem to be solved is a nonconvex function, gradient descent-based algorithms are not guaranteed to find the global optimal solution. To obtain the global optimal solution, adaptive step size, stochastic gradient descent, and momentum methods are often used [24–26].

Algorithm 5.1: Pseudo code for FMEEF-GD algorithm

Input: Step size μ, Weighting coefficient λ, Kernel bandwidths σ_1, σ_2, Sample number N

Output: Weight vector $\boldsymbol{w}$ of the filter

$m \leftarrow 1$, $\boldsymbol{w}(1) \leftarrow \boldsymbol{0}$, $\xi_1 \leftarrow \mu\lambda / \sigma_1^2$, $\xi_2 \leftarrow \mu(1-\lambda)/\sigma_2^2$

while $m \leq M$ **do**

  $e(m) \leftarrow d(m) - \boldsymbol{w}(m)^{\mathrm{T}} \boldsymbol{x}(m)$

  $g_1(e(m)) \leftarrow k_{\sigma_1}(e(m))e(m)\boldsymbol{x}(m)$

  if $m \leq N$ **then**

    $\boldsymbol{w}(m+1) \leftarrow \boldsymbol{w}(m) + \mu\lambda g_1(e(m))$

  else

    $\bar{e}(m) \leftarrow \mathrm{mean}\{e(m-N), \cdots, e(m-1)\}$

    $g_2(e(m)-\bar{e}(m)) \leftarrow k_{\sigma_2}(e(m)-\bar{e}(m))(e(m)-\bar{e}(m))(\boldsymbol{u}(m)-\bar{\boldsymbol{u}}(m))$

    $\boldsymbol{w}(m+1) \leftarrow \boldsymbol{w}(m) + \xi_1 g_1(e(m)) + \xi_2 g_2(e(m))$

  end

  $m \leftarrow m+1$

end

return $\boldsymbol{w}$

5.4 Performance Analysis

In this section, we conduct an analysis of the transient and steady-state performance for the FMEEF-GD algorithm in terms of both the mean and mean-square. In addition, we provide complexity analysis for the FMEEF-GD algorithm. We first list our assumptions:

A1 The input vector $\boldsymbol{x}(m)$ obeys the zero-mean multivariate independent Gaussian distribution with covariance matrix denoted as $\boldsymbol{R}_{xx} = \mathrm{E}\left[\boldsymbol{x}(m)\boldsymbol{x}^{\mathrm{T}}(m)\right]$ [27].
A2 The noise $v(m)$ follows the zero-mean independent, identically distributed process with variance $\mathrm{E}\left[v^2(m)\right] = \sigma_v^2$ [28].
A3 The input vector $\boldsymbol{u}(m)$, the noise $v(m)$, the MGF of $e^2(m)$ and that of $\overline{e}^2(m)$ are mutually independent [16].

In our analysis, we denote the weight-error vector $\hat{\boldsymbol{w}}(m)$ as

$$\hat{\boldsymbol{w}}(m) = \boldsymbol{w}(m) - \boldsymbol{w}_{\mathrm{opt}} \tag{5.18}$$

Subsequently, the error $e(m)$ and the error mean $\overline{e}(m)$ can be written as

$$e(m) = d(m) - \boldsymbol{w}^{\mathrm{T}}(m)\boldsymbol{x}(m) = v(m) - \hat{\boldsymbol{w}}^{\mathrm{T}}(m)\boldsymbol{x}(m) \tag{5.19}$$

$$\overline{e}(m) = \frac{1}{N}\sum_{i=m-N}^{m-1}\left[v(i) - \hat{\boldsymbol{w}}^{\mathrm{T}}(m)\boldsymbol{x}(i)\right] \tag{5.20}$$

Subtracting $\boldsymbol{w}_{\mathrm{opt}}$ from both sides of (5.14), we have

$$\begin{aligned}\hat{\boldsymbol{w}}(m+1) &= \hat{\boldsymbol{w}}(m) + \xi_1 k_{\sigma_1}(e(m))e(m)\boldsymbol{x}(m)\\ &\quad + \xi_2 k_{\sigma_2}(e(m)-\overline{e}(m))(e(m)-\overline{e}(m))(\boldsymbol{x}(m)-\overline{\boldsymbol{x}}(m))\end{aligned} \tag{5.21}$$

Substituting (5.19) and (5.20) into (5.21) yields

$$\begin{aligned}\hat{\boldsymbol{w}}(m+1) &= \hat{\boldsymbol{w}}(m) - \xi_1 k_{\sigma_1}(e(m))\boldsymbol{x}(m)\boldsymbol{x}^{\mathrm{T}}(m)\hat{\boldsymbol{w}}(m)\\ &\quad + \xi_1 k_{\sigma_1}(e(m))v(m)\boldsymbol{x}(m) + \xi_2 k_{\sigma_2}(e(m)-\overline{e}(m))v(m)\boldsymbol{x}(m)\\ &\quad - \xi_2 k_{\sigma_2}(e(m)-\overline{e}(m))\boldsymbol{x}(m)\boldsymbol{x}^{T}(m)\hat{\boldsymbol{w}}(m) - \xi_2 k_{\sigma_2}(e(m)-\overline{e}(m))\overline{v}(m)\boldsymbol{x}(m)\\ &\quad + \xi_2 k_{\sigma_2}(e(m)-\overline{e}(m))\overline{d}(m)\boldsymbol{x}(m) - \xi_2 k_{\sigma_2}(e(m)-\overline{e}(m))v(m)\overline{\boldsymbol{x}}(m)\\ &\quad + \xi_2 k_{\sigma_2}(e(m)-\overline{e}(m))\overline{\boldsymbol{x}}(m)\boldsymbol{x}^{T}(m)\hat{\boldsymbol{w}}(m) + \xi_2 k_{\sigma_2}(e(m)-\overline{e}(m))\overline{v}(m)\overline{\boldsymbol{x}}(m)\\ &\quad - \xi_2 k_{\sigma_2}(e(m)-\overline{e}(m))\overline{d}(m)\overline{\boldsymbol{x}}(m)\end{aligned} \tag{5.22}$$

where

$$\overline{v}(m) = \frac{1}{N}\sum_{i=m-N}^{m-1} v(i) \tag{5.23}$$

$$\overline{d}(m) = \frac{1}{N}\sum_{i=m-N}^{m-1} \hat{\boldsymbol{w}}^{\mathrm{T}}(i)\boldsymbol{x}(i) \tag{5.24}$$

Since the weight-error vector $\hat{\boldsymbol{w}}(m)$ depends on the input vector before time m, we assume independence between $\hat{\boldsymbol{w}}(m)$ and the input vector $\boldsymbol{x}(m)$. Based on this assumption, the subsequent statistical analysis can be significantly simplified.

5.4.1 *Mean Behavior*

There are some important results that can be utilized based on statistics. First, according to the central limit theorem and Assumptions A1 and A2, we observe that the measurement noise mean $\overline{v}(m)$ and the input mean $\overline{\boldsymbol{x}}(m)$ obey zero-mean Gaussian distribution with variance σ_v^2/N and covariance $\boldsymbol{R}_{xx}/N$, respectively. Second, from Assumption A3, we see that $\boldsymbol{x}(m), \overline{\boldsymbol{x}}(m), v(m)$ and $\overline{v}(m)$ are independent of each other. Given a small step size, we can use time m to approximate the weight-error vector for the first N times. Taking mathematical expectation on both sides of (5.22) and utilizing Assumptions A1-A3, we obtain

$$\begin{aligned}\mathrm{E}[\hat{\boldsymbol{w}}(m+1)] \approx\ & \mathrm{E}[\hat{\boldsymbol{w}}(m)] - \xi_1 \mathrm{E}\big[k_{\sigma_1}(e(m))\big]\boldsymbol{R}_{xx}\mathrm{E}[\hat{\boldsymbol{w}}(m)] \\ & - \xi_2 \mathrm{E}\big[k_{\sigma_2}(e(m) - \overline{e}(m))\big]\boldsymbol{R}_{xx}\mathrm{E}[\hat{\boldsymbol{w}}(m)] \\ & - \xi_2 \mathrm{E}\big[k_{\sigma_2}(e(m) - \overline{e}(m))\big]\boldsymbol{R}_{xx}\mathrm{E}[\hat{\boldsymbol{w}}(m)]/N \end{aligned} \tag{5.25}$$

To find the behavior of the FMEEF-GD algorithm, we need to calculate the expected values of the nonlinear terms $k_{\sigma_1}(e(m))$ and $k_{\sigma_2}(e(m) - \overline{e}(m))$. In the literature [29], the statistical average approximation method is used to estimate the expected values of nonlinear terms, while in fact, these terms vary with weights and cannot be considered as deterministic values when the step size is large. Inspired by the MGF, the expected values of the nonlinear terms can be calculated by computing the MGF of the square of the random variable, i.e.,

$$M_X(s) = \mathrm{E}\left[e^{sx^2}\right] = \int_{-\infty}^{+\infty} e^{sx^2} p(x)\mathrm{d}x \tag{5.26}$$

where $X = x^2$, s is the variable, and $p(x)$ is the probability density function of x. We can obtain the MGF of the same form as the Gaussian kernel function by setting $s = -1/\sigma^2$. Since the mean vector of the input vector $\boldsymbol{x}(m)$ and the mean of the noise $v(m)$ are assumed to be zero, the expected values are:

$$M_{z_1(m)}(s) \approx M_{V(m)}(s)M_{X(m)}(s) \tag{5.27}$$

$$M_{z_2(m)}(s) \approx M_{V(m)}(s)M_{X(m)}(s)M_{\overline{V}(m)}(s)M_{\overline{X}(m)}(s) \tag{5.28}$$

where $z_1(m) = e^2(m)$, $z_2(m) = (e(m) - \overline{e}(m))^2$, $U(m) = \hat{\boldsymbol{w}}^{\mathrm{T}}(m)\boldsymbol{x}(m)\boldsymbol{x}^{\mathrm{T}}(m)\hat{\boldsymbol{w}}(m)$, $V(m) = v^2(m)$, $\overline{V}(m) = \overline{v}^2(m)$, and $\overline{U}(m) = \overline{d}(m)\overline{d}(m)$. These approximations imply that the kernel bandwidths satisfy $\sigma_1 \geq 1$ and $\sigma_2 \geq 1$ so that the solution of the cross-term expectation is disregarded. With a sufficiently small step size, we can replace the weight-error vector of the previous N time with the weight-error vector of time m, that is $\overline{U}(m) \approx \hat{\boldsymbol{w}}^{\mathrm{T}}(m)\overline{\boldsymbol{x}}(m)\overline{\boldsymbol{x}}^{\mathrm{T}}(m)\hat{\boldsymbol{w}}(m)$. Therefore, the expected values of the nonlinear terms is written as

$$\mathrm{E}\big[k_{\sigma_1}(e(m))\big] = M_{z_1(m)}(-1/\sigma_1^2) \tag{5.29}$$

$$\mathrm{E}\big[k_{\sigma_2}(e(m) - \overline{e}(m))\big] = M_{z_2(m)}(-1/\sigma_2^2) \tag{5.30}$$

In addition, the MGF of the quadratic form of the random vector obeying Gaussian distribution is expressed as [30, pp.100]:

$$M_{X(m)}(s) = (1 - 2s\mathrm{Tr}(\boldsymbol{R}_{xx}\boldsymbol{Q}(m)))^{-\frac{1}{2}} \tag{5.31}$$

$$M_{\overline{X}(m)}(s) = (1 - 2s\mathrm{Tr}(\boldsymbol{R}_{xx}\boldsymbol{Q}(m))/N)^{-\frac{1}{2}} \tag{5.32}$$

where $\boldsymbol{Q}(m) = \mathrm{E}\big[\hat{\boldsymbol{w}}(m)\hat{\boldsymbol{w}}^{\mathrm{T}}(m)\big]$ and $\mathrm{Tr}(\cdot)$ represents trace of the matrix. Similarly, the noise-mean-squared MGF is evaluated as:

$$M_{\overline{V}(m)}(s) = (1 - 2s\sigma_v^2/N)^{-\frac{1}{2}} \tag{5.33}$$

To facilitate obtaining the behavior of the FMEEF-GD algorithm, the noise-squared MGFs for different distributions are given in Table 2.1.

Based on the above development, (5.25) is simplified as

$$\mathrm{E}[\hat{\boldsymbol{w}}(m+1)] = \left[\boldsymbol{I}_L - \eta_{m_1}\boldsymbol{R}_{xx} - \eta_{m_2}\boldsymbol{R}_{xx}\frac{\eta_{m_2}\boldsymbol{R}_{xx}}{N}\right]\mathrm{E}[\hat{\boldsymbol{w}}(m)] \tag{5.34}$$

where $\boldsymbol{I}_L$ refers to the $L \times L$ identity matrix, and

$$\eta_{m_1} = \xi_1 M_{z_1(m)}\left(-\frac{1}{2\sigma_1^2}\right) \tag{5.35}$$

$$\eta_{m_2} = \xi_2 M_{z_2(m)}\left(-\frac{1}{2\sigma_2^2}\right) \tag{5.36}$$

Therefore, the transient weight of the FMEEF-GD algorithm in the mean sense is determined from

$$\begin{aligned} \mathrm{E}[\boldsymbol{w}(m+1)] = & \left[\eta_{m_1}\boldsymbol{R}_{xx} + \eta_{m_2}\boldsymbol{R}_{xx} + \frac{\eta_{m_2}\boldsymbol{R}_{xx}}{N}\right]\boldsymbol{w}_{\text{opt}} \\ & + \left[\boldsymbol{I}_L - \eta_{m_1}\boldsymbol{R}_{xx} - \eta_{m_2}\boldsymbol{R}_{xx}\frac{\eta_{m_2}\boldsymbol{R}_{xx}}{N}\right]\mathrm{E}[\boldsymbol{w}(m)] \end{aligned} \tag{5.37}$$

Based on (5.34), the FMEEF-GD algorithm (5.14) converges asymptotically for any initial condition when the step size meets the following equation:

$$0 < \mu < \frac{2N\sigma_1^2\sigma_2^2}{\lambda_{\max}\left(\left(\lambda\sigma_2^2 NM_{z_1(m)}\left(-\frac{1}{2\sigma_1^2}\right) + (1-\lambda)(1+N)\sigma_1^2 M_{z_2(m)}\left(-\frac{1}{2\sigma_2^2}\right)\right)\boldsymbol{R}_{xx}\right)} \tag{5.38}$$

where $\lambda_{\max}(\cdot)$ indicates the largest eigenvalue of the matrix.

5.4.2 Mean-Square Behavior

The mean-square deviation (MSD) is described as

$$\mathrm{MSD}(m) = \mathrm{Tr}(\boldsymbol{Q}(m)) = \mathrm{E}\left[\|\hat{\boldsymbol{w}}(m)\|^2\right] \tag{5.39}$$

Multiplying both sides of Eq. (5.22) by its transpose, and subsequently taking the expectation on both sides while considering Assumptions A1-A3, results in:

$$\begin{aligned} \boldsymbol{Q}(m+1) \approx\ & \boldsymbol{Q}(m) + \eta_{m_1}^2\sigma_v^2\boldsymbol{R}_{xx} + \eta_{m_2}^2\sigma_v^2\boldsymbol{R}_{xx} \\ & + 2\eta_{m_2}^2\sigma_v^2\boldsymbol{R}_{xx}/N + \eta_{m_2}^2\sigma_v^2\boldsymbol{R}_{xx}/N^2 + 2\eta_{m_1}\eta_{m_2}\sigma_v^2\boldsymbol{R}_{xx} \\ & - (\eta_{m_1} + \eta_{m_2} + \eta_{m_2}/N)[\boldsymbol{R}_{xx}\boldsymbol{Q}(m) + \boldsymbol{Q}(m)\boldsymbol{R}_{xx}] \\ & + (\eta_{m_1} + \eta_{m_2})^2\mathrm{E}\left[\boldsymbol{x}(m)\boldsymbol{x}^T(m)\hat{\boldsymbol{w}}(m)\hat{\boldsymbol{w}}^T(m)\boldsymbol{x}(m)\boldsymbol{x}^T(m)\right] \\ & + 2(\eta_{m_1}\eta_{m_2} + \eta_{m_2}^2)\mathrm{E}\left[\overline{d}(m)\overline{\boldsymbol{x}}(m)\hat{\boldsymbol{w}}^T(m)\boldsymbol{x}(m)\boldsymbol{x}^T(m)\right] \\ & + \eta_{m_2}^2\mathrm{E}\left[\overline{\boldsymbol{x}}(m)\boldsymbol{x}^T(m)\hat{\boldsymbol{w}}(m)\hat{\boldsymbol{w}}^T(m)\boldsymbol{x}(m)\overline{\boldsymbol{x}}^T(m)\right] \\ & + \eta_{m_2}^2\mathrm{E}\left[\overline{d}(m)\boldsymbol{x}(m)\boldsymbol{x}^T(m)\overline{d}(m)\right] \\ & + \eta_{m_2}^2\mathrm{E}\left[\overline{d}(m)\overline{\boldsymbol{x}}(m)\overline{\boldsymbol{x}}^T(m)\overline{d}(m)\right] \end{aligned} \tag{5.40}$$

We can decompose the fourth-order moments in (5.40) utilizing the Gaussian moment factorization theorem [31] into

$$\mathrm{E}\left[\boldsymbol{x}(m)\boldsymbol{x}^{\mathrm{T}}(m)\hat{\boldsymbol{w}}(m)\hat{\boldsymbol{w}}^{\mathrm{T}}(m)\boldsymbol{x}(m)\boldsymbol{x}^{\mathrm{T}}(m)\right] = 2\boldsymbol{R}_{xx}\boldsymbol{Q}(m)\boldsymbol{R}_{xx} + \boldsymbol{R}_{xx}\mathrm{Tr}(\boldsymbol{R}_{xx}\boldsymbol{Q}(m)) \tag{5.41}$$

When the step size is small enough, the remaining fourth-order moments in (5.40) are approximated as

$$\mathrm{E}\left[\overline{d}(m)\overline{\boldsymbol{x}}(m)\hat{\boldsymbol{w}}^{T}(m)\boldsymbol{x}(m)\boldsymbol{x}^{T}(m)\right] \approx \frac{\boldsymbol{R}_{xx}\boldsymbol{Q}(m)\boldsymbol{R}_{xx}}{N} \tag{5.42}$$

$$\mathrm{E}\left[\overline{\boldsymbol{x}}(m)\boldsymbol{x}^{\mathrm{T}}(m)\hat{\boldsymbol{w}}(m)\hat{\boldsymbol{w}}^{\mathrm{T}}(m)\boldsymbol{x}(m)\overline{\boldsymbol{x}}^{\mathrm{T}}(m)\right] \approx \boldsymbol{R}_{xx}\mathrm{Tr}(\boldsymbol{R}_{xx}\boldsymbol{Q}(m))/N \tag{5.43}$$

$$\mathrm{E}\left[\overline{d}(m)\boldsymbol{x}(m)\boldsymbol{x}^{\mathrm{T}}(m)\overline{d}(m)\right] \approx \boldsymbol{R}_{xx}\mathrm{Tr}(\boldsymbol{R}_{xx}\boldsymbol{Q}(m))/N \tag{5.44}$$

$$\mathrm{E}\left[\overline{d}(m)\overline{\boldsymbol{x}}(m)\overline{\boldsymbol{x}}^{\mathrm{T}}(m)\overline{d}(m)\right] \approx 2\boldsymbol{R}_{xx}\boldsymbol{Q}(m)\boldsymbol{R}_{xx}/N^2 + \boldsymbol{R}_{xx}\mathrm{Tr}(\boldsymbol{R}_{xx}\boldsymbol{Q}(m))/N^2 \tag{5.45}$$

Substituting (5.40–5.45) into (5.39) yields

$$\begin{aligned}
\boldsymbol{Q}(m+1) \approx{} & \boldsymbol{Q}(m) + [(\eta_{m_1} + \eta_{m_2})^2 + 2\eta_{m_2}^2/N + \eta_{m_2}^2/N^2]\sigma_v^2\boldsymbol{R}_{xx} \\
& - (\eta_{m_1} + \eta_{m_2} + \eta_{m_2}/N)[\boldsymbol{R}_{xx}\boldsymbol{Q}(m) + \boldsymbol{Q}(m)\boldsymbol{R}_{xx}] \\
& + 2\left[(\eta_{m_1} + \eta_{m_2})^2 + \eta_{m_2}^2/N^2 + (\eta_{m_1}\eta_{m_2} + \eta_{m_2}^2)/N\right]\boldsymbol{R}_{xx}\boldsymbol{Q}(m)\boldsymbol{R}_{xx} \\
& + \left[(\eta_{m_1} + \eta_{m_2})^2 + \eta_{m_2}^2/N^2 + 2\eta_{m_2}^2/N\right]\boldsymbol{R}_{xx}\mathrm{Tr}(\boldsymbol{R}_{xx}\boldsymbol{Q}(m))
\end{aligned} \tag{5.46}$$

To derive the transient behavior of $\boldsymbol{Q}(m)$, the following equalities are employed:

$$\mathrm{Tr}(\mathbf{AB}) = \left(\mathrm{vec}\left(\boldsymbol{A}^{\mathrm{T}}\right)\right)^{\mathrm{T}}\mathrm{vec}(\mathbf{B}) \tag{5.47}$$

$$\mathrm{vec}(\mathbf{ABC}) = \left[\mathbf{C}^{T} \otimes \boldsymbol{A}\right]\mathrm{vec}(\mathbf{B}) \tag{5.48}$$

Vectorizing both sides of (5.46) and using (5.47) and (5.48), we have

$$\begin{aligned}
\mathrm{vec}(\boldsymbol{Q}(m+1)) ={} & \mathrm{vec}(\boldsymbol{Q}(m)) + [(\eta_{m_1} + \eta_{m_2})^2 + 2\eta_{m_2}^2/N + \eta_{m_2}^2/N^2]\sigma_v^2\mathrm{vec}(\boldsymbol{R}_{xx}) \\
& - (\eta_{m_1} + \eta_{m_2} + \eta_{m_2}/N)[\boldsymbol{I} \otimes \boldsymbol{R}_{xx}]\mathrm{vec}(\boldsymbol{Q}(m)) \\
& - (\eta_{m_1} + \eta_{m_2} + \eta_{m_2}/N)\left[\boldsymbol{R}_{xx}^{T} \otimes \boldsymbol{I}\right]\mathrm{vec}(\boldsymbol{Q}(m)) \\
& + 2\left[(\eta_{m_1} + \eta_{m_2})^2 + \eta_{m_2}^2/N^2 + (\eta_{m_1}\eta_{m_2} + \eta_{m_2}^2)/N\right] \\
& \left[\boldsymbol{R}_{xx}^{\mathrm{T}} \otimes \boldsymbol{R}_{xx}\right]\mathrm{vec}(\boldsymbol{Q}(m)) + \left[(\eta_{m_1} + \eta_{m_2})^2 + \eta_{m_2}^2/N^2 + 2\eta_{m_2}^2/N\right] \\
& \mathrm{vec}(\boldsymbol{R}_{xx})\mathrm{vec}\left(\boldsymbol{R}_{xx}^{\mathrm{T}}\right)^{\mathrm{T}}\mathrm{vec}(\boldsymbol{Q}(m))
\end{aligned} \tag{5.49}$$

According to (5.49), the MSD value at time m is calculated using $\mathrm{vec}(\boldsymbol{I}_L)^{\mathrm{T}}\mathrm{vec}(\boldsymbol{Q}(m))$.

5.4.3 Mean-Square Steady-State Behavior

Assume that the algorithm reaches steady state when $m \to \infty$, such that $\boldsymbol{Q}(m)$ meets

$$\lim_{m\to\infty} \boldsymbol{Q}(m) = \lim_{m\to\infty} \boldsymbol{Q}(m+1) \tag{5.50}$$

Substituting (5.50) into (5.49), we can simplify the steady-state equation of the FMEEF-GD algorithm to

$$\begin{aligned}
&\left(\eta_{\infty_1} + \eta_{\infty_2} + \eta_{\infty_2}/N\right)[\boldsymbol{I} \otimes \boldsymbol{R}_{xx}]\mathrm{vec}(\boldsymbol{Q}(\infty)) \\
&+\left(\eta_{\infty_1} + \eta_{\infty_2} + \eta_{\infty_2}/N\right)\left[\boldsymbol{R}_{xx}^{\mathrm{T}} \otimes \boldsymbol{I}\right]\mathrm{vec}(\boldsymbol{Q}(\infty)) \\
&-2\left[\left(\eta_{\infty_1} + \eta_{\infty_2}\right)^2 + \eta_{\infty_2}^2/N^2 + \left(\eta_{\infty_1}\eta_{\infty_2} + \eta_{\infty_2}^2\right)/N\right]\left[\boldsymbol{R}_{xx}^{\mathrm{T}} \otimes \boldsymbol{R}_{xx}\right]\mathrm{vec}(\boldsymbol{Q}(\infty)) \\
&-\left[\left(\eta_{\infty_1} + \eta_{\infty_2}\right)^2 + \eta_{\infty_2}^2/N^2 + 2\eta_{\infty_2}^2/N\right]\mathrm{vec}(\boldsymbol{R}_{xx})\mathrm{vec}\left(\boldsymbol{R}_{xx}^{\mathrm{T}}\right)^{\mathrm{T}}\mathrm{vec}(\boldsymbol{Q}(\infty)) \\
&= \left[\left(\eta_{\infty_1} + \eta_{\infty_2}\right)^2 + 2\eta_{\infty_2}^2/N + \eta_{\infty_2}^2/N^2\right]\sigma_v^2\mathrm{vec}(\boldsymbol{R}_{xx})
\end{aligned} \tag{5.51}$$

Solving (5.51) results in

$$\mathrm{vec}(\boldsymbol{Q}(\infty)) = \boldsymbol{P}^{-1}[(\eta_{\infty_1} + \eta_{\infty_2})^2 + 2\eta_{\infty_2}^2/N + \eta_{\infty_2}^2/N^2]\sigma_v^2\mathrm{vec}(\boldsymbol{R}_{xx}) \tag{5.52}$$

where

$$\begin{aligned}
\boldsymbol{P} = {} & (\eta_{\infty_1} + \eta_{\infty_2} + \eta_{\infty_2}/N)[\boldsymbol{I} \otimes \boldsymbol{R}_{xx}] + (\eta_{\infty_1} + \eta_{\infty_2} + \eta_{\infty_2}/N)\left[\boldsymbol{R}_{xx}^{\mathrm{T}} \otimes \boldsymbol{I}\right] \\
& - 2\left[(\eta_{\infty_1} + \eta_{\infty_2})^2 + \eta_{\infty_2}^2/N^2 + (\eta_{\infty_1}\eta_{\infty_2} + \eta_{\infty_2}^2)/N\right]\left[\boldsymbol{R}_{xx}^{\mathrm{T}} \otimes \boldsymbol{R}_{xx}\right] \\
& - \left[(\eta_{\infty_1} + \eta_{\infty_2})^2 + \eta_{\infty_2}^2/N^2 + 2\eta_{\infty_2}^2/N\right]\mathrm{vec}(\boldsymbol{R}_{xx})\mathrm{vec}\left(\boldsymbol{R}_{xx}^{\mathrm{T}}\right)^{\mathrm{T}}
\end{aligned} \tag{5.53}$$

In the steady state, $\mathrm{Tr}(\boldsymbol{R}_{xx}\boldsymbol{Q}(\infty)) \ll \sigma_v^2$ and $\mathrm{Tr}(\boldsymbol{R}_{xx}\boldsymbol{Q}(\infty)) \approx 0$ can be assumed to hold, and substituting it into (5.31) and (5.32) yields $M_{X(\infty)} = M_{\overline{X}(\infty)} = 1$. Therefore, η_{∞_1} and η_{∞_2} are obtained as

$$\eta_{\infty_1} = \xi_1 M_{V(m)}\left(-\frac{1}{2\sigma_1^2}\right) \tag{5.54}$$

$$\eta_{\infty_2} = \xi_2 M_{V(m)}\left(-\frac{1}{2\sigma_2^2}\right)M_{\overline{V}(m)}\left(-\frac{1}{2\sigma_2^2}\right) \tag{5.55}$$

Therefore, the steady-state MSD can be derived by combining (5.52–5.55). It is worth noting that the Taylor expansion was not used in our analysis to approximate the nonlinear term in the weight update Eq. (5.14), thus leading to a better approximation of the steady-state MSD than the Taylor expansion method [32].

Table 5.1 Computational complexities of four algorithms

Algorithms	Multiplication	Addition	Exponentiation
LMS	$2L + 1$	$2L$	0
MC	$2L + 8$	$2L$	1
MEEF-GD	$2NL + 5N + L + 13$	$3NL + N + L + 1$	$N + 1$
FMEEF-GD	$3L + 19$	$4L + N + 1$	2

5.4.4 *Complexity Analysis*

In this subsection, we explore the computational complexities of LMS [17], MC, MEEF-GD and FMEFF-GD algorithms. The numbers of multiplication, addition and exponentiation operations per iteration are tabulated in Table 5.1.

It is seen that the complexities of all four algorithms appear to increase with the filter length L. For MEEF-GD algorithm, the sample number N is the decisive factor in the computation. Since the FMEEF-GD algorithm replaces the a posteriori errors with the a priori errors, it is only necessary to calculate the mean value of the errors instead of repeatedly calculating the a posteriori errors at time m. As a result, the numbers of the multiplication and exponentiation of our method are independent of the sample number N while the addition number shows a linear relationship with the sample number.

5.5 Simulation Results

In this section, Monte Carlo simulations are conducted under diverse noise environments to validate the theoretical results. Additionally, we assess the FMEEF-GD algorithm in comparison to several existing robust adaptive filtering algorithms. The FMEEF-GD algorithm.

is finally deployed in AEC.

Consider the system identification scenario in which the noisy output $d(m)$ is obtained by a linear model (5.7). The system vector $\boldsymbol{w}_{\text{opt}}$ is created randomly, and the default initial weight vector for the algorithm is the null vector. The input signal $\boldsymbol{x}(m)$ obeys a zero-mean white Gaussian processes. The MSD in decibel (dB) is employed as a performance metric, defined as $10\log_{10}\left(\|\hat{\mathbf{w}}(m)\|^2\right)$. In addition, the curves presented in simulations are acquired by averaging the results of 50 independent Monte Carlo trials.

5.5.1 Gaussian Noise Case

First, we examine the mean behavior evolution of all weight values of FMEEF-GD algorithm under Gaussian noise environment. The step size $\mu = 0.01$, the kernel bandwidths $\sigma_1 = 2$ and $\sigma_2 = 2$, the filter length $L = 5$, the sample number $N = 30$, and the variance of noise $\sigma_v^2 = 0.01$ are considered. Figure 5.4 depicts the variation of each weight mean of the FMEEF-GD algorithm.

As shown in Fig. 5.4, the theoretical curve resulted from (5.37) matches very well with the simulation results.

Further, we assess both the steady-state and transient behaviors of MSD for FMEEF-GD algorithm. In our simulation, we consider a filter length of 20. Figure 6.8 show the learning curves of Monte Carlo simulations conducted with various values of step size μ, noise variance σ_v^2, kernel bandwidths and weighting coefficient λ, as well as the theoretical transient and steady-state behaviors derived from (5.49) and (5.52), respectively. In the figures, dotted line refers to the simulation, and the dashed and solid lines depict the steady-state and transient behaviors. It is observed that our theoretical model predicts well the steady-state and transient behaviors of the FMEEF-GD algorithm. Figures 5.5, 5.7, and 5.8 illustrate that the convergence speed of the FMEEF-GD algorithm is influenced by the step size, kernel bandwidth, and weighting coefficient. In this configuration, the FMEEF-GD algorithm converges slower and the steady-state MSD is lower when the step size μ is reduced, or the kernel bandwidths σ_1 and σ_2 are increased, or the weighting coefficient λ is increased. In addition, while the noise variance σ_v^2 decreasing leads to a decrease in the steady-state MSD, it does not influence the convergence rate of the algorithm, as shown in Fig. 5.6.

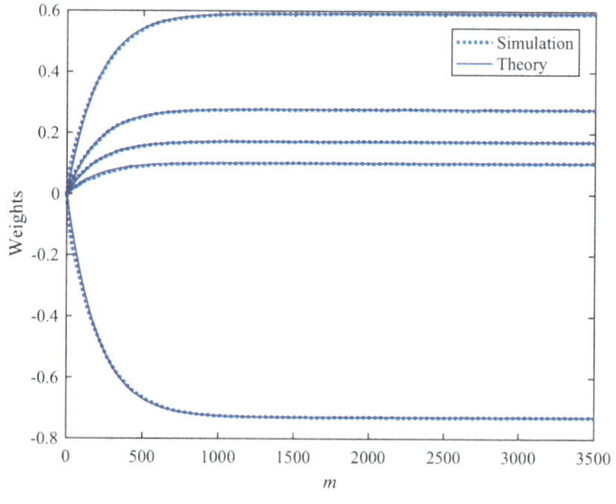

Fig. 5.4 Simulated and theoretical learning curves of each weight mean for FMEEF-GD algorithm

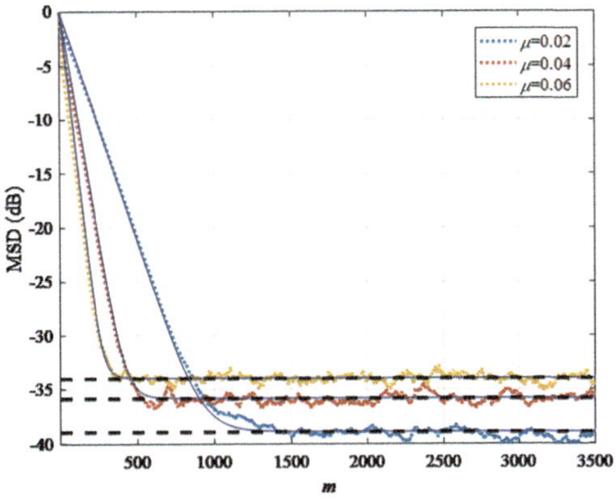

Fig. 5.5 MSDs of FMEEF-GD algorithm versus iteration number in Gaussian noise at different μ

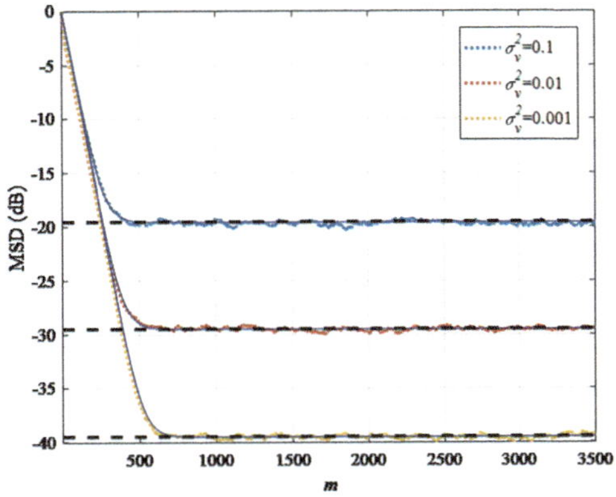

Fig. 5.6 MSDs of FMEEF-GD algorithm versus iteration number in Gaussian noise at different σ_v^2

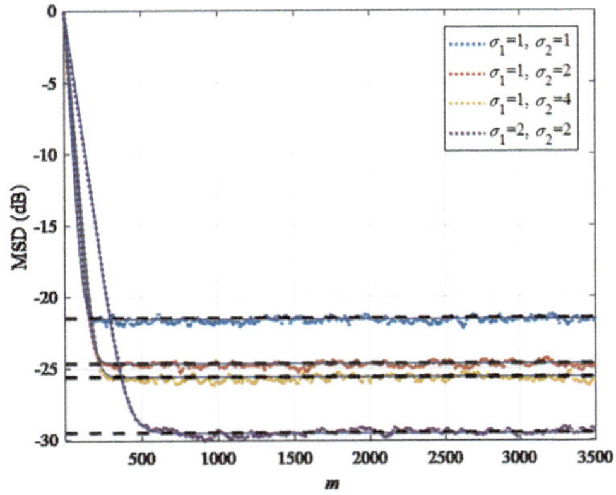

Fig. 5.7 MSDs of FMEEF-GD algorithm versus iteration number in Gaussian noise at different σ_1 and σ_2

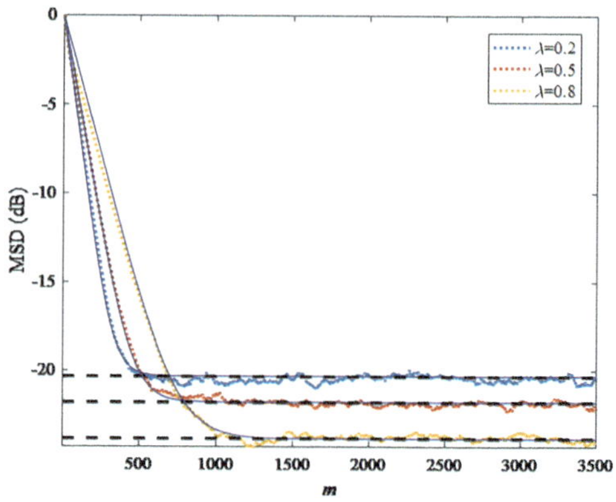

Fig. 5.8 MSDs of FMEEF-GD algorithm versus iteration number in Gaussian noise at different λ

5.5.2 *Non-Gaussian Noise Case*

Now we verify the theoretical model when the noise obeys binomial distribution over $\{-\gamma, \gamma\}$ with the probability mass function $Pr(v = \gamma) = 0.5$ and uniform distribution over the interval $[-a, a]$. In the simulation, the parameters of the FMEEF-GD algorithm remain unchanged. Figures 5.9 and 5.10 show the steady-state and transient MSD behaviors under various conditions of the step sizes and noise variances when the measurement noise follows binomial distribution. When the measurement noise follows uniform distribution, the evolution of the steady-state and transient behaviors of the MSD under various conditions of the step sizes and noise variances are plotted in Figs. 5.11 and 5.12. Again, we observe that the theoretical curves of MSD are in good agreement with both steady-state and transient simulation results. As observed in Figs. 5.10 and 5.12, decreasing the step size μ results in a slower convergence speed and a decrease in the steady-state MSD for the FMEEF-GD algorithm. In addition, as shown in Figs. 5.9 and 5.11, the steady-state MSD reduces as the noise variance σ_v^2 decreases.

5.5.3 *Algorithms Comparison*

First, we compare the convergence curves of MSD and the runtimes of LMS [17], MC [16], MEEF-GD [11] and FMEEF-GD algorithms. The measurement noise $v(m)$ adopts two common noise models, namely asymmetric Gaussian mixture distribution and alpha-state distribution, where the asymmetric Gaussian mixture distribution with $0.9N(0, 1) + 0.1N(1, 9)$ [29], and the alpha-stable distribution with the tail parameter 1.4, variance parameter 0.1 and mean parameter 0.5. The kernel bandwidth in MC algorithm is $\sigma = 2$. The kernel bandwidths in MEEF-GD and FMEEF-GD algorithms are $\sigma_1 = 2$ and $\sigma_2 = 2$ with weighting factor $\lambda = 0.43$. The filter length is $L = 20$. We set the step sizes of all algorithms such that their initial convergence

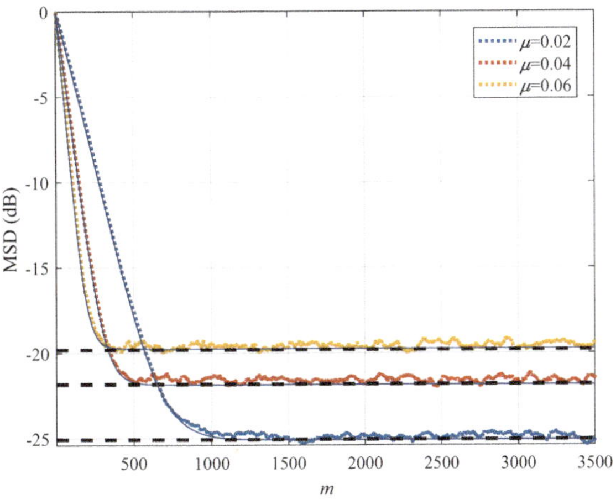

Fig. 5.9 MSDs of FMEEF-GD algorithm versus iteration number in binomial noise at different μ

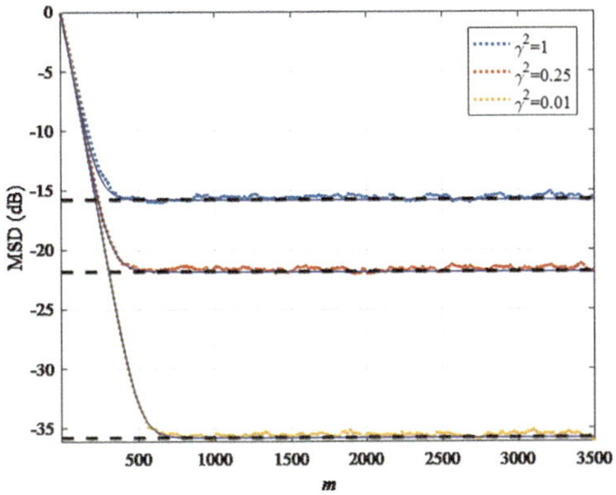

Fig. 5.10 MSDs of FMEEF-GD algorithm versus iteration number in binomial noise at different γ^2

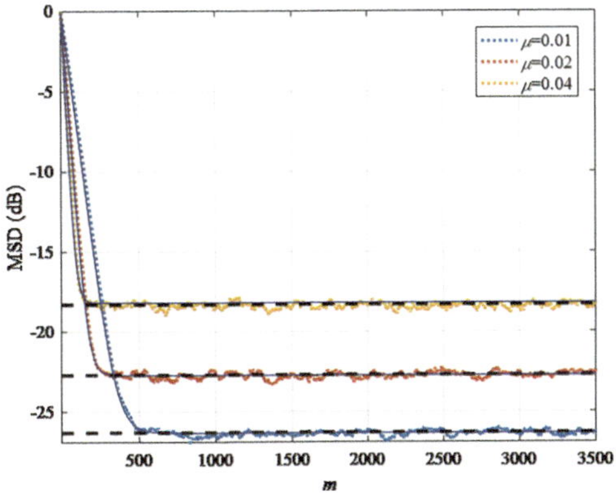

Fig. 5.11 MSDs of FMEEF-GD algorithm versus iteration number in uniform noise at different μ

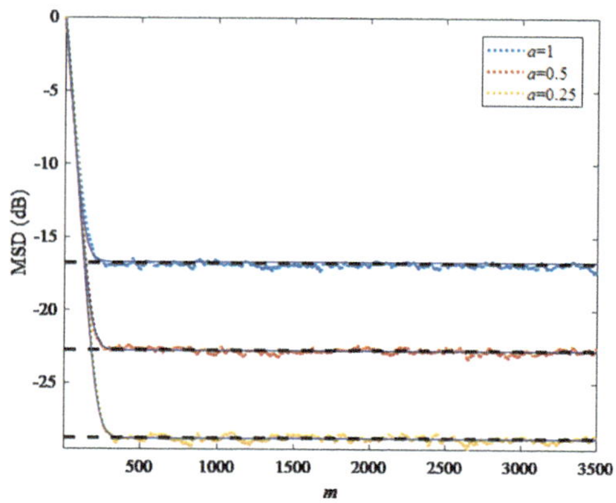

Fig. 5.12 MSDs of FMEEF-GD algorithm versus iteration number in uniform noise at different a

rates are the same. The MSD learning curves of the LMS, MC, MEEF-GD, and FMEEF-GD algorithms in different noise environments are shown in Figs. 5.13 and 5.14, respectively, where the number of samples is $N = 10$. In addition, we compare their runtimes at different values of N using MATLAB 2016b with AMD Ryzen 7 5700U, 1.80 GHz CPU, under asymmetric Gaussian mixture distribution noise environment. Table 5.2 show the results the total runtime with the required iteration number being 3500 and the repetition number being 50. It is seen from Table 5.2 that the FMEEF-GD algorithm effectively improves the drawback of the MEEF-GD algorithm that the runtime increases with the sample number. It is worth noting that the MEEF-GD and FMEEF-GD algorithms have similar results to the LMS algorithm in symmetric Gaussian mixed noise environments, while the MC algorithm performs poorly as shown in Fig. 5.13. On the other hand, the MC, MEEF-GD, and FMEEF-GD algorithms perform more robustly compared to the LMS algorithm in an alpha-stable noise environment as shown in Fig. 5.14. Overall, the FMEEF-GD algorithm and the MEEF-GD algorithm have the same steady-state MSD, but the FMEEF-GD algorithm significantly outperforms the MEEF-GD algorithm in computational complexity.

Second, we assess the performance of the LMS, MC, MEEF-GD and FMEEF-GD algorithms in wind signal amplitude prediction. The wind data used in this comparison were collected at the Institute of Industrial Science, University of Tokyo, as documented in [33]. In the simulation, we apply a moving average filter with a window size of 30 to preprocess the raw data. To conduct the evaluation, we divide the east–west dataset into two segments. The first 3000 samples are designated as the training dataset, while the remaining 2000 samples are assigned to the test dataset. With these datasets in place, the performance of the four algorithms in predicting wind signal amplitudes is evaluated and compared. The kernel bandwidths and weighting factors for each algorithm are kept the same values as in the previous configuration. However, we fine-tune the step size of each algorithm to ensure the best prediction accuracy for the wind signal amplitude prediction task. Figure 5.15 displays the prediction results of wind signals of the high region by the four algorithms. As can

Fig. 5.13 MSDs of MC, MEEF-GD and FMEEF-GD algorithms versus iteration number in asymmetric Gaussian mixture noise environment

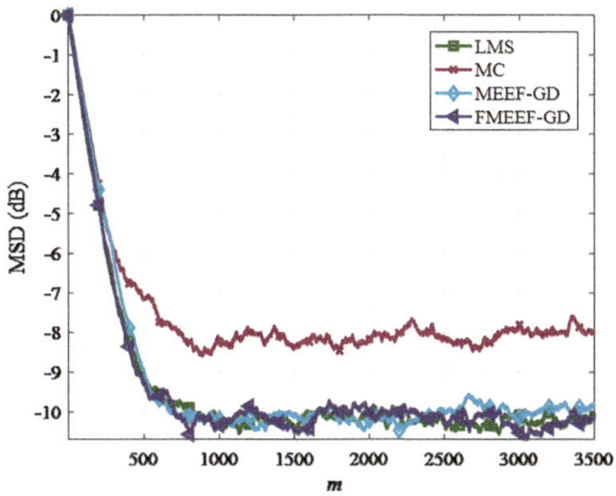

Fig. 5.14 MSDs of MC, MEEF-GD and FMEEF-GD algorithms versus iteration number in alpha-stable noise environment

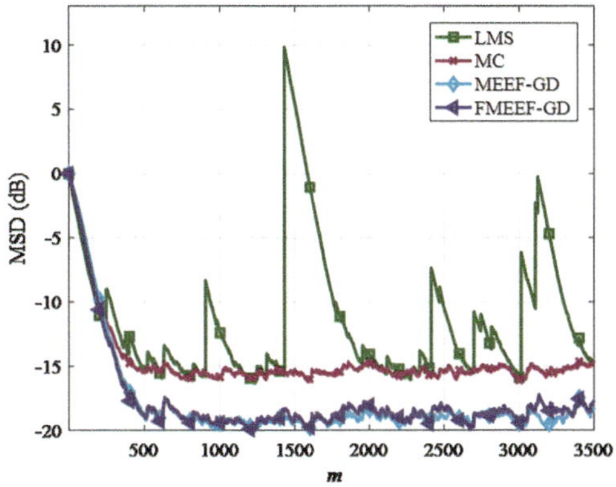

Table 5.2 Runtime of different algorithms

Algorithms	Runtime(s)			
	$N = 10$	$N = 20$	$N = 30$	$N = 40$
LMS	0.72	0.72	0.73	0.73
MC	0.73	0.72	0.74	0.73
MEEF-GD	7.37	12.16	16.93	21.92
FMEEF-GD	1.79	1.83	1.92	1.94

be seen in Fig. 5.15, all four algorithms are able to accurately predict the wind signal magnitude. To further compare the prediction accuracies of the different algorithms for the four algorithms in different regions, the predicted mean square errors for the low, medium, and high regions are summarized in Table 5.3. As anticipated, both the FMEEF-GD and MEEF-GD algorithms exhibit identical accuracy levels in predicting the low and medium regions. While the accuracy of the FMEEF-GD algorithm in the high region may be slightly inferior to that of the MEEF-GD algorithm,

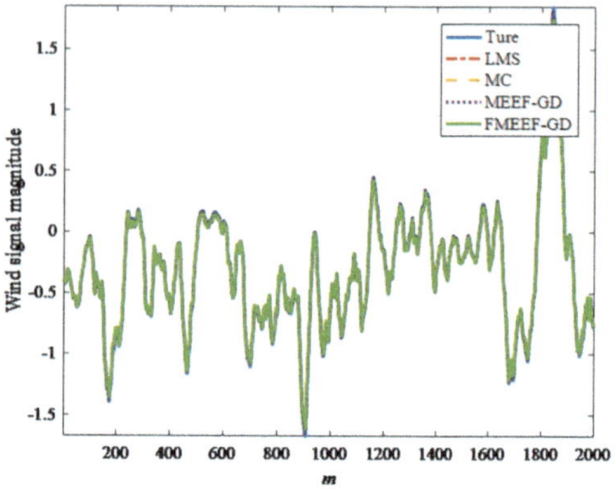

Fig. 5.15 Wind signal prediction comparison of high region

Table 5.3 Wind signal prediction mean square error of the low, medium and high regions

Algorithms	LMS	MC	MEEF-GD	FMEEF-GD
Low	0.0002	0.0002	0.0001	0.0001
Medium	0.0007	0.0004	0.0004	0.0004
High	0.0015	0.0015	0.0005	0.0011

this result is entirely acceptable considering the significantly lower computational complexity. Overall, the FMEEF-GD algorithm has some advantages in wind signal prediction.

5.5.4 *Acoustic Echo Cancellation*

Finally, the proposed algorithm is applied to both single-talk and double-talk scenarios in AEC. The normalized MSD (NMSD), commonly employed as the metric in AEC, is defined as $10\log_{10}E\left[\|\hat{\mathbf{w}}(m)\|^2/\|\mathbf{w}_{\text{opt}}\|^2\right]$. The impulse response of the echo path $\mathbf{w}_{opt}$ with length $L = 256$ is given in Fig. 5.16. The sampling frequency of the speech signal is 8 kHz and the sampling time is 20 s [34, 35]. During the iteration process, we emulate the echo path evolution by shifting to the right by 12 bits, thereby assessing the tracking performance of the algorithms. The noise has two signal sources, namely Gaussian noise with mean 0 and variance 0.1 and alpha-stable noise with mean 0.1, variance 0.1 and tail parameter 1.4. The parameters used in the manuscript are $\mu = 0.002$ for LMS algorithm, $\mu = 0.002$, $\sigma = 1$ for MC algorithm, $\mu = 0.003$, $\sigma_1 = 1$, $\sigma_2 = 1$, $\lambda = 0.5$, $N = 10$ for MEEF-GD algorithm, and $\mu = 0.003$, $\sigma_1 = 1$, $\sigma_2 = 1$, $\lambda = 0.5$, $N = 10$ for FMEEF-GD algorithm. The NMSD curves of the four algorithms are given in Fig. 5.17 for the single-talk and double-talk cases in alpha-stable noise. It is evident from Fig. 5.17, the LMS algorithm performs worst in non-Gaussian noise environment due to its susceptibility to the impulsive noise. The MEEF-GD and FMEEF-GD algorithms outperform the MC algorithm with the same initial convergence rate. Since the complexity of the FMEEF-GD algorithm is significantly lower than that of the MEEF-GD algorithm, the FMEEF-GD algorithm offers a remarkable advantage in AEC.

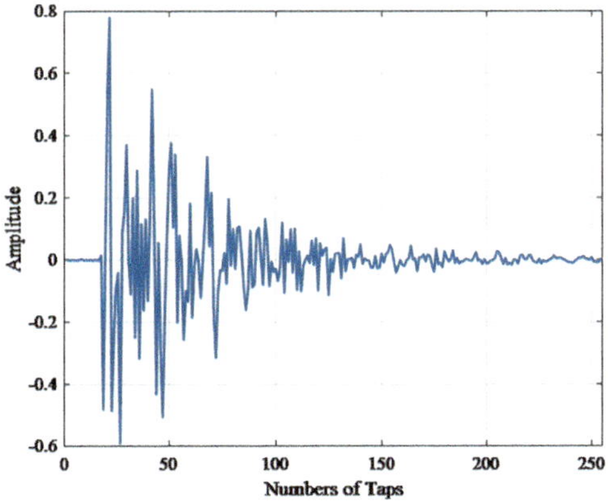

Fig. 5.16 Impulse response of the echo path

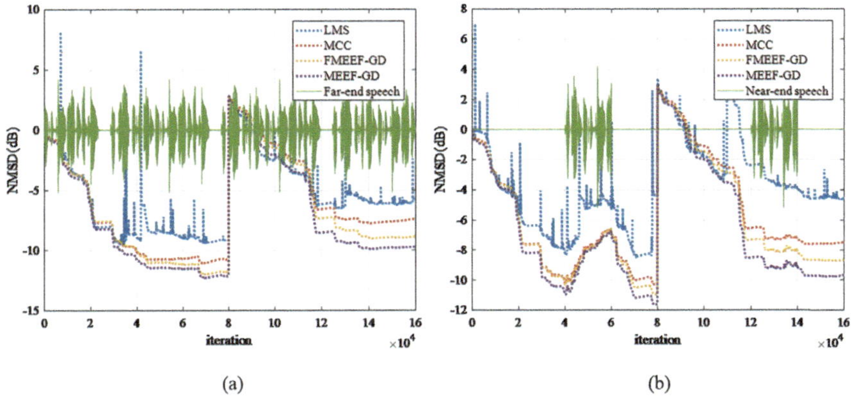

Fig. 5.17 NMSD convergence curves for the algorithms are compared in scenarios with alpha-stable noise: **a** single-talk and **b** double-talk

5.6 Summary

In this study, we address the computational complexity associated with the MEEF criterion. We propose an FMEEF criterion and derive a corresponding adaptive filtering algorithm based on GD method. Moreover, the steady-state and transient behaviors of the FMEEF-GD algorithm under various noise distributions are discussed. Simulation results show excellent agreement between the simulated curves and theoretical calculation when $\sigma_1 > 1$ and $\sigma_2 > 1$. It is worth noting that the proposed algorithm exhibits the similar transient and steady-state behaviors as the MEEF-GD algorithm, but the complexity of the FMEEF-GD algorithm is almost independent of the number of error samples, which greatly reduces the computational requirement.

References

1. H. Radmanesh, M. Hajiabadi, Recursive maximum correntropy learning algorithm with adaptive kernel size. IEEE Trans. Circuits Syst. II Express Briefs **65**(7), 958–962 (2018)
2. W. Liu, P.P. Pokharel, J.C. Principe, Correntropy: Properties and applications in non-Gaussian signal processing. IEEE Trans. Signal Process. **55**(11), 5286–5298 (2007)
3. Y. Zheng, B. Chen, S. Wang, W. Wang, W. Qin, Mixture correntropy-based kernel extreme learning machines. IEEE Trans. Neural Netw. Learn. Syst. **33**(2), 811–825 (2022)
4. D. Erdogmus, J.C. Principe, An error-entropy minimization algorithm for supervised training of nonlinear adaptive systems. IEEE Trans. Signal Process. **50**(7), 1780–1786 (2002)
5. K. Xiong, H.H.C. Iu, S. Wang, Kernel correntropy conjugate gradient algorithms based on half-quadratic optimization. IEEE Trans. Cybern. **51**(11), 5497–5510 (2021)
6. H. Song, D. Ding, H. Dong, Q.-L. Han, Distributed maximum correntropy filtering for stochastic nonlinear systems under deception attacks. IEEE Trans. Cybern. **52**(5), 3733–3744 (2022)
7. M. Li, Z. Jing, H. Leung, Robust minimum error entropy based cubature information filter with non-Gaussian measurement noise. IEEE Signal Process. Lett. **28**, 349–353 (2021)
8. B. Chen, L. Xing, N. Zheng, J.C. Principe, Quantized minimum error entropy criterion. IEEE Trans. Neural Netw. Learn. Syst. **30**(5), 1370–1380 (2019)
9. W. Liu, P. P. Pokharel, J. C. Principe, Error entropy, correntropy and m-estimation, in *2006 16th IEEE Signal Processing Society Workshop on Machine Learning for Signal Processing* (2006), pp. 179–184
10. Y. Xie, Y. Li, Y. Gu, J. Cao, B. Chen, Fixed-point minimum error entropy with fiducial points. IEEE Trans. Signal Process. **68**, 3824–3833 (2020)
11. R.J. Bessa, V. Miranda, J. Gama, Entropy and correntropy against minimum square error in offline and online three-day ahead wind power forecasting. IEEE Trans. Power Syst. **24**(4), 1657–1666 (2009)
12. R. Mitra, G. Kaddoum, G. Dahman, G. Poitau, Error analysis of localization based on minimum-error entropy with fiducial points. IEEE Commun. Lett. **25**(4), 1187–1191 (2021)
13. L. Dang, B. Chen, Y. Huang, Y. Zhang, H. Zhao, Cubature kalman filter under minimum error entropy with fiducial points for ins/GPS integration. IEEE/CAA J. Automatica Sinica **9**(3), 450–465 (2022)
14. A.R. Heravi, A new robust fixed-point algorithm and its convergence analysis. J. fixed point theory appl. **19**, 3191–3215 (2017)
15. Y. Zheng, S. Wang, B. Chen, Quantized minimum error entropy with fiducial points for robust regression. Neural Netw. **168**, 405–418 (2023)

16. X. Hou, H. Zhao, X. Long, An innovative transient analysis of adaptive filter with maximum correntropy criterion. IEEE Signal Process. Lett. **29**, 1689–1693 (2022)
17. A. H. Sayed, *Fundamentals of Adaptive Filtering* (Wiley, 2003)
18. T. Hu, Q. Wu, D.-X. Zhou, Convergence of gradient descent for minimum error entropy principle in linear regression. IEEE Trans. Signal Process. **64**(24), 6571–6579 (2016)
19. B. Chen, L. Xing, B. Xu, H. Zhao, J.C. Principe, Insights into the robustness of minimum error entropy estimation. IEEE Trans. Neural Netw. Learn. Syst. **29**(3), 731–737 (2018)
20. W. A. Martins, P. S. R. Diniz, Y.-F. Huang, On the normalized minimum error-entropy adaptive algorithm: cost function and update recursion, in *2010 First IEEE Latin American Symposium on Circuits and Systems (LASCAS)* (2010), pp. 140–143
21. S. Han, S. Rao, K.-H. Jeong, J. C. Principe, A normalized minimum error entropy stochastic algorithm, in *2006 IEEE International Conference on Acoustics Speech and Signal Processing Proceedings*, vol. 5 (2006), pp. 1–5
22. D. Erdogmus, K. Hild, J.C. Principe, Online entropy manipulation: stochastic information gradient. IEEE Signal Process. Lett. **10**(8), 242–245 (2003)
23. S. Han, S. Rao, D. Erdogmus, K.-H. Jeong, J. Principe, A minimum-error entropy criterion with self-adjusting step-size (mee-sas). Signal Process. **87**(11), 2733–2745 (2007)
24. R. Huang, Y. Qin, K. Liu, G. Yuan, Biased stochastic conjugate gradient algorithm with adaptive step size for nonconvex problems. Expert Syst. Appl. **238**, 121556 (2024)
25. Y. Lei, T. Hu, G. Li, K. Tang, Stochastic gradient descent for nonconvex learning without bounded gradient assumptions. IEEE Trans. Neural Netw. Learn. Syst. **31**(10), 4394–4400 (2020)
26. Y. Liang, J. Liu, D. Xu, Stochastic momentum methods for non-convex learning without bounded assumptions. Neural Netw. **165**, 830–845 (2023)
27. L. Li, H. Zhao, A robust total least mean m-estimate adaptive algorithm for impulsive noise suppression. IEEE Trans. Circuits Syst. II Express Briefs **67**(4), 800–804 (2020)
28. X. Long, H. Zhao, X. Hou, A novel combinatoric correntropy algorithm: Properties and its performance analysis. IEEE Trans. Circuits Syst. II Express Briefs **69**(12), 5184–5188 (2022)
29. G. Qian, J. Mei, H.H.C. Iu, S. Wang, Fixed-point maximum total complex correntropy algorithm for adaptive filter. IEEE Trans. Signal Process. **69**, 2188–2202 (2021)
30. J. Omura, T. Kailath, *Some useful probability distributions*, Technical Report 7050-6, The Stanford Electronics Laboratories, Stanford University, Stanford, CA (1965)
31. V. Mathews, Z. Xie, A stochastic gradient adaptive filter with gradient adaptive step size. IEEE Trans. Signal Process. **41**(6), 2075–2087 (1993)
32. G. Qian, X. Ning, S. Wang, Recursive constrained maximum correntropy criterion algorithm for adaptive filtering. IEEE Trans. Circuits Syst. II Express Briefs **67**(10), 2229–2233 (2020)
33. D. Mandic, S. Javidi, S. Goh, A. Kuh, K. Aihara, Complex-valued prediction of wind profile using augmented complex statistics. Renewable Energy **34**(1), 196–201 (2009)
34. S. Lv, H. Zhao, W. Xu, Robust affine projection krsmpl adaptive filtering algorithm and its application. IEEE Trans. Circuits Syst. II Express Briefs **71**(3), 1506–1510 (2024)
35. K. Dogancay, O. Tanrikulu, Adaptive filtering algorithms with selective partial updates. IEEE Trans. Circ. Syst. II: Analog Dig. Signal Process. **48**(8), 762–769 (2001)

Chapter 6
Asymmetric Correntropy Criterion

6.1 Introduction

With the development of adaptive filtering techniques, the adaptive algorithms being researched and designed are now in full bloom in the field of processing signals of different data dimensions, such as LMS, NLMS algorithms in the real domain for processing one-dimensional data signals [1, 2]; CLMS, WL-CLMS algorithms in the complex-valued domain for processing two-dimensional data [3, 4]; Geometric-algebraic LMS and quaternion LMS algorithms for processing high-dimensional data [5–7], etc. Among those, LMS-type algorithms have received wide attention since they have been developed and proposed due to its simple design and easy implementation. However, in practical applications (for example, the echo phenomenon commonly occurs in devices such as vehicle hands-free communication systems and video conferencing systems [8–10]), the algorithms are affected by natural noise, i.e., impulsive noise regardless of the dimension of the data signal being processed.

It is precisely in this case that the performance of LMS-type algorithms is severely degraded. Considering practical engineering application problems, it is crucial to study and design algorithms that are robust to impulsive noise. The most popular one that has been studied and proposed is the MCC algorithm based on the maximum correntropy criterion [11–21]. The correntropy based on information theoretic learning (ITL) is used as a measure of similarity, which is sensitive to outliers when the kernel width is small value because correntropy contains higher order information about the error signal [15]. Thus, the designed algorithms based on the MCC criterion have strong robustness against non-Gaussian noises [22].

The algorithms mentioned above are proposed for symmetrically distributed noise environments, in other words, the kernel functions used by the algorithms are symmetric Gaussian kernels. However, there are still many situations where asymmetric noise occurs when processing data, such as financial analysis [23], image processing [24], and so on. Therefore, an asymmetric and robust loss function was proposed in [25]. Then, in the field of signal processing, the maximum asymmetric

H. Zhao et al., *Information Theoretic Learning-Based Filter*,
https://doi.org/10.1007/978-3-032-29623-8_6

correntropy algorithm (MACC) is proposed to fill the gap in the field of processing asymmetric signals as well as noise [26]. Subsequently, the generalized maximum asymmetric correntropy algorithm (GMACC) was proposed in order to improve the performance of the MACC algorithm [27]. Immediately after that, the asymmetric correntropy was better promoted, such as the finite asymmetric generalized Gaussian mixture model learning is used to improve the effect of infrared object detection in [28]. In [29], an affine projection asymmetric correntropy algorithm was proposed. In [30], the asymmetric correntropy was applied to the complex-valued domain for solving asymmetric complex-valued domain signals. It is worth mentioning that the current research on asymmetric correntropy algorithms is still relatively scarce, and the proposed algorithms are in single-kernel form, whose correntropy lacks flexibility and adaptability. Therefore, in order to optimize the performance of the algorithms as well as to enhance the generality of the algorithms, it is significant to study the MACC algorithms in multi-kernel form.

In an asymmetric noise environment, robust algorithms designed based on symmetric kernels are no longer applicable. Non-Gaussian kernels, on the other hand, can utilize Gaussian kernel functions with different bandwidths to separately handle positive and negative estimation errors, thus better adapting to asymmetric noise environments. Therefore, this chapter proposes a dual-kernel asymmetric correntropy (DK-AC) by leveraging the characteristics of Gaussian and power exponential kernels. Based on the DK-AC, a dual-kernel generalized maximum asymmetric correntropy (DK-GMAC) criterion is constructed. A robust filtering algorithm, namely the dual-kernel generalized maximum asymmetric correntropy criterion (DK-GMACC), is then developed according to the DK-GMAC criterion. The stability conditions of the algorithm are provided, and the steady-state performance of the algorithm is analyzed in detail. The correctness of the theoretical analysis is verified in the experimental phase. Finally, the algorithm's performance is compared and tested in application scenarios such as SI, echo channel estimation and echo cancellation, and WP. The results demonstrate the superiority of the proposed DK-GMACC algorithm under different noise distributions and in various application scenarios.

6.2 Asymmetric Correntropy Criterion

Based on the ITL, the similarity measure of two variables can be expressed using correntropy. Given two random variables c_1 and c_2 with $c_1, c_2 \in \mathbb{R}$, the correntropy between them is defined as [15]:

$$S^G_{C,\sigma}(c_1, c_2) = E[\kappa(c_1, c_2)] = \int \kappa(c_1, c_2) dF_{c_1 c_2}(c_1, c_2), \tag{6.1}$$

where $\kappa(c_1, c_2)$ is the kernel function, $E[\cdot]$ is the expectation operator, $F_{c_1c_2}(c_1, c_2)$ denotes the joint distribution function of (c_1, c_2). The $\kappa(c_1, c_2)$ is in general expressed using a Gaussian kernel function, i.e.,

$$\kappa(c_1, c_2) = \kappa_\sigma^G(c_1, c_2) = \frac{1}{\sqrt{2\pi}\sigma} \exp\left[-\frac{(c_1 - c_2)^2}{2\sigma^2}\right], \tag{6.2}$$

where $\kappa_\sigma^G(c_1, c_2)$ represents the Gaussian kernel function, σ denotes the kernel width.

Here considering a linear model, the error variable C_i is defined as the difference between $c_{1,i}$ and $c_{2,i}$, i.e., $C_i = c_{1,i} - c_{2,i}$. Since $\{c_{1,i}, c_{2,i}\}_{i=1}^N$ in the data (c_1, c_2) is discrete and the joint distribution $F_{c_1c_2}(c_1, c_2)$ is unknown, the correntropy cannot be derived by direct computation. Therefore, in practice, $S_{C,\sigma}^G(c_1, c_2)$ is generally estimated using a finite sample dataset $\{c_{1,i}, c_{2,i}\}_{i=1}^N$, that is

$$\hat{S}_{C,\sigma}^G(c_1, c_2) = \frac{1}{N}\sum_{i=1}^{N} \exp\left[-\frac{\left(c_{1,i} - c_{2,i}\right)^2}{2\sigma_1^2}\right] = \frac{1}{N}\sum_{i=1}^{N} \kappa_{C,\sigma}^G\left(c_{1,i} - c_{2,i}\right). \tag{6.3}$$

In signal data processing, there are

$$\hat{S}_{e,\sigma}^G(d, y) = \frac{1}{N}\sum_{m=1}^{N} \kappa_{e,\sigma}^G(d(m) - y(m)) = \frac{1}{N}\sum_{m=1}^{N} \kappa_{e,\sigma}^G(e(m)), \tag{6.4}$$

where the error $e(m)$ is defined as the difference between the desired signal $d(m) = \boldsymbol{w}_{opt}^T \boldsymbol{x}(m) + v(m)$ and the filter output $y(m) = \boldsymbol{w}^T \boldsymbol{x}(m)$, i.e., $e(m) = d(m) - y(m)$, $\boldsymbol{w}_{opt}$ is the vector of weights need to be estimated and v_i is the measurement noise. $\boldsymbol{w} = [w_1, w_2, \ldots, w_M]^T \in \mathbb{R}^{M\times 1}$ represents the weight vector, $\boldsymbol{x}(m) = [x_m, x_{m-1}, \ldots, x_{m-M+1}]^T \in \mathbb{R}^{M\times 1}$ is the input signal, $(\cdot)^T$ denotes the transpose operation, M is the order of the filter.

For some asymmetric noise, the MCC will no longer be applicable. In this case, the MCC is replaced by using the MACC, which replaces the Gaussian kernel in the MCC with an asymmetric Gaussian kernel with different kernel widths to deal with the different distributions on both sides of the peaks of the asymmetric noise. The expression for the asymmetric correntropy is defined as

$$G_{\sigma_+,\sigma_-}^A(e(m)) = \begin{cases} \exp\left(-\dfrac{e^2(m)}{2\sigma_+^2}\right), \text{ if } e(m) \geq 0 \\ \exp\left(-\dfrac{e^2(m)}{2\sigma_-^2}\right), \text{ if } e(m) < 0 \end{cases}, \tag{6.5}$$

where σ_+, σ_- are the kernel widths. And the generalized asymmetric Gaussian kernel is

$$G^{\mathrm{GA}}_{\sigma_+,\sigma_-}(e(m))=\begin{cases}\exp(-|e(m)/\sigma_+|^{\gamma}), \text{ if } e(m)\geq 0\\ \exp(-|e(m)/\sigma_-|^{\gamma}), \text{ if } e(m)< 0\end{cases}. \tag{6.6}$$

where γ is the shape parameter.

6.3 Adaptive Filtering Based on Asymmetric Correntropy Criterion

6.3.1 *The Cost Function*

Since the Gaussian kernel function has smoothness and strict positive definiteness; the power exponential kernel function has heavy-tailedness and can better handle outliers. We will combine these two kernel functions as a new kernel function. In addition, considering the universality and applicability of asymmetric correntropy, we use a generalized form. The new asymmetric correntropy proposed in this chapter is called the dual-kernel generalized asymmetric correntropy (DK-GAC). It is defined as

$$S^{\mathrm{DK}}_{C,\sigma}(e_i) = E\left[\kappa^{\mathrm{DK}}_{e,\sigma}(e(m))\right], \tag{6.7}$$

where

$$\kappa^{\mathrm{DK}}_{e,\sigma}(e(m))=\begin{cases}\exp\left(-\left(\alpha_1\dfrac{|e(m)|^{\gamma}}{2\sigma^2_{1,+}}+\alpha_2\dfrac{|e(m)|^{\gamma-1}}{\sigma_{2,+}}\right)\right), \text{ if } e(m)\geq 0\\ \exp\left(-\left(\alpha_1\dfrac{|e(m)|^{\gamma}}{2\sigma^2_{1,-}}+\alpha_2\dfrac{|e(m)|^{\gamma-1}}{\sigma_{2,-}}\right)\right), \text{ if } e(m)< 0\end{cases}, \tag{6.8}$$

where α_1, α_2 are the proportionality parameter which represents the occupancy of the two kernel functions with $\alpha_2 = 1 - \alpha_1$. $\sigma_{1,+}$, $\sigma_{1,-}$, $\sigma_{2,+}$, and $\sigma_{2,-}$ denote the kernel widths.

Remark 6.1 Formally, our proposed dual-kernel correntropy is essentially obtained by multiplying two different kernel functions, which is different from the traditionally cumulative form of multi-kernel correntropy [22]. Therefore, we provide a new way of thinking to improve the performance of correntropy algorithms compared with the traditional multi-kernel correntropy obtained by accumulating single-kernel correntropy.

The cost function based on DK-GAC is defined as

$$J_{\mathrm{DK-GMACC}} = \kappa^{DK}_{e,\sigma}(0) - E\left[\kappa^{DK}_{e,\sigma}(e(m))\right], \tag{6.9}$$

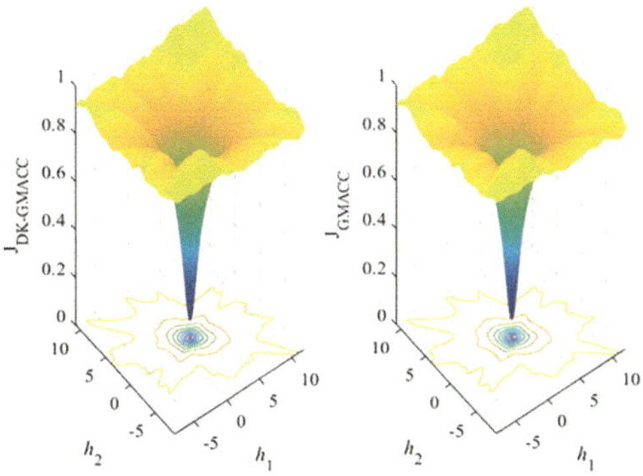

Fig. 6.1 Performance surfaces of DK-GMACC and GMACC criteria

where $E[\kappa_{e,\sigma}^{DK}(e(m))]$ is equal to the average value of the kernel function any time in the case of discrete sample inputs,

$$E\left[\kappa_{e,\sigma}^{DK}(e(m))\right] \approx \frac{1}{N}\sum_{m=1}^{N}\kappa_{e,\sigma}^{DK}(e(m)). \tag{6.10}$$

Therefore, Eq. (6.9) can be re-expressed as

$$J_{DK-GMACC} = 1 - \frac{1}{N}\sum_{m=1}^{N}\kappa_{e,\sigma}^{DK}(e(m)). \tag{6.11}$$

In order to visualize the advantages of $J_{DK-GMACC}$, we give the performance surfaces and gradient differences (grad$\{J_{DK-GMACC}\}$ − grad$\{J_{GMACC}\}$) plots of $J_{DK-GMACC}$ and J_{GMACC} with respect to h_1 and h_2. In the experiments, a two-tap finite impulse response filter with $\boldsymbol{w}_{opt} = \boldsymbol{h} = [h_1, h_2]^T = [2, 1]^T$ is used [26, 31, 32], $\gamma = 2$, $\alpha_1 = 0.998$, $\sigma_+ = \sigma_{1,+} = 3$, $\sigma_- = \sigma_{1,-} = \sigma_{2,-} = 1.2$, $\sigma_{2,+} = 2.4$. In a Gaussian white noise environment with variance 0.1, the comparison results are shown in Figs. 6.1 and 6.2. Figure 6.1 illustrates that J_{GMACC} and $J_{DK-GMACC}$ have similar performance. And Fig. 6.2 demonstrates that $J_{DK-GMACC}$ has a larger gradient compared to J_{GMACC}, which also validates that the proposed algorithms in the experimental part has a faster convergence speed.

6.3.2 Deviation of the DK-GMACC Algorithm

For the system identification problem, the optimal weight vector of the system can be obtained by adaptive filtering method. This optimization problem is equal to maximizes the cost function [33], the objective function for solving $\boldsymbol{w}$ is thus

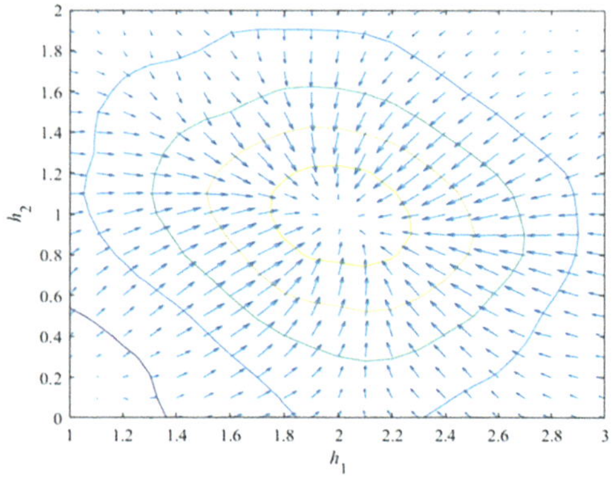

Fig. 6.2 Contour and gradient difference between DK-GMACC and GMACC criteria

$$\boldsymbol{w} = \underset{\mathbf{w}\in\mathbb{R}^M}{\arg\max}\left\{E\left[\kappa_{e,\sigma}^{DK}(e(m))\right]\right\} = \frac{1}{N}\sum_{m=1}^{N}\kappa_{e,\sigma}^{DK}(e(m)). \tag{6.12}$$

To facilitate the derivation, we reorganize (6.8) as

$$\phi_{\text{DK-GAC}}(e(m)) = \begin{cases} \exp\left(-\left(\alpha_1\beta_{1,1}|e(m)|^{\gamma} + \alpha_2\beta_{2,1}|e(m)|^{\gamma-1}\right)\right), \text{ if } e(m) \geq 0 \\ \exp\left(-\left(\alpha_1\beta_{1,2}|e(m)|^{\gamma} + \alpha_2\beta_{2,2}|e(m)|^{\gamma-1}\right)\right), \text{ if } e(m) < 0 \end{cases}, \tag{6.13}$$

where $\beta_{1,1} = 1/2\sigma_{1,+}^2$, $\beta_{1,2} = 1/2\sigma_{1,-}^2$, $\beta_{2,1} = 1/\sigma_{2,+}$, and $\beta_{2,2} = 1/\sigma_{2,-}$.

Using Eq. (6.13), a stochastic gradient algorithm called in this work the DK-GMACC can be derived as

$$\begin{aligned} \mathbf{w}(m+1) &= \mathbf{w}(m) + \mu\frac{\partial\phi_{\text{DK-GAC}}(e(m))}{\partial\mathbf{w}(m)} \\ &= \mathbf{w}(m) - \mu\Phi_{\text{DK-GAC}}(e(m))|e(m)|^{\gamma-1}\frac{\partial e(m)}{\partial\mathbf{w}(m)} \\ &= \mathbf{w}(m) + \mu\Phi_{\text{DK-GAC}}(e(m))|e(m)|^{\gamma-1}\boldsymbol{x}(m), \end{aligned} \tag{6.14}$$

where μ is the step-size and

$$\Phi_{\text{DK-GAC}}(e(m)) = \begin{cases} \left(\alpha_1\beta_{1,1} + \dfrac{\alpha_2\beta_{2,1}}{|e(m)|}\right)f_1(e(m)), \text{ if } e(m) \geq 0 \\ \left(\alpha_1\beta_{1,2} + \dfrac{\alpha_2\beta_{2,2}}{|e(m)|}\right)f_2(e(m)), \text{ if } e(m) < 0 \end{cases}, \tag{6.15}$$

where $f_1(e(m)) = \exp(-(\alpha_1\beta_{1,1}|e(m)|^{\gamma} + \alpha_2\beta_{2,1}|e(m)|^{\gamma-1}))$, $f_2(e(m)) = \exp(-(\alpha_1\beta_{1,2}|e(m)|^{\gamma} + \alpha_2\beta_{2,2}|e(m)|^{\gamma-1}))$.

Simplifying (6.14) again, we have

$$\boldsymbol{w}(m+1) = \boldsymbol{w}(m) + \eta_{\sigma_+\sigma_-}(m)\phi_{\text{DK-GAC}}(e(m))|e(m)|^{\gamma-1}\boldsymbol{x}(m), \tag{6.16}$$

where

$$\eta_{\sigma_+\sigma_-}(m) = \begin{cases} \mu\left(\alpha_1\beta_{1,1} + \dfrac{\alpha_2\beta_{2,1}}{|e(m)|+\delta}\right), & \text{if } e(m) \geq 0 \\ \mu\left(\alpha_1\beta_{1,2} + \dfrac{\alpha_2\beta_{2,2}}{|e(m)|+\delta}\right), & \text{if } e(m) < 0 \end{cases}, \tag{6.17}$$

which denotes the step-size, δ is a normal number designed to prevent division by zero with $0 < \delta \ll 1$.

6.4 Properties

It is worth noting that the DK-GMACC algorithm has several features.

Property 6.1 *When $\alpha_1 = 1$, the DK-GMACC algorithm becomes the GMACC algorithm* [28], *the weight update equation of the GMACC algorithm can be expressed as*

$$\boldsymbol{w}(m+1) = \boldsymbol{w}(m) + \mu\Phi_{\text{GMACC}}(e(m))e(m)\boldsymbol{x}(m), \tag{6.18}$$

where

$$\Phi_{\text{GMACC}}(e(m)) = \begin{cases} \exp(-\beta_{2,1}|e(m)|^{\gamma})|e(m)|^{\gamma-2}, \text{if } e(m) \geq 0 \\ \exp(-\beta_{2,2}|e(m)|^{\gamma})|e(m)|^{\gamma-2}, \text{if } e(m) < 0 \end{cases}. \tag{6.19}$$

Property 6.2 *When $\alpha_1 = 1$ and $\gamma = 2$, the DK-GMACC algorithm becomes the MACC algorithm* [26], *which can be written as*

$$\boldsymbol{w}(m+1) = \boldsymbol{w}(m) + \mu\Phi_{\text{MACC}}(e(m))e(m)\boldsymbol{x}(m), \tag{6.20}$$

where

$$\Phi_{\text{MACC}}(e(m)) = \begin{cases} \exp(-\beta_{2,1}|e(m)|^{2}), \text{if } e(m) \geq 0 \\ \exp(-\beta_{2,2}|e(m)|^{2}), \text{if } e(m) < 0 \end{cases}. \tag{6.21}$$

Property 6.3 *When $\sigma_{1,+} = \sigma_{1,-}$ and $\sigma_{2,+} = \sigma_{2,-}$, the DK-GMACC algorithm becomes the C-MCC algorithm* [16]. *When $\sigma_{1,+} = \sigma_{1,-}$, $\sigma_{2,+} = \sigma_{2,-}$, and $\alpha_1 = 1$, the DK-GMACC algorithm becomes the GMCC algorithm* [14]. *When $\sigma_{1,+} = \sigma_{1,-}$, $\sigma_{2,+} = \sigma_{2,-}$, $\alpha_1 = 1$, and $\gamma = 2$, the DK-GMACC algorithm becomes the MCC algorithm* [17]. *In addition, by employing the properties of the GMCC, when $\sigma_{1,+} = \sigma_{1,-} \to \infty$ and $\alpha_1 = 1$, the DK-GMACC algorithm becomes the LMP algorithm*

[34], *and when* $\sigma_{1,+} = \sigma_{1,-} \to \infty$, $\gamma = 2$, *and* $\alpha_1 = 1$, *the DK-GMACC algorithm becomes the LMS algorithm* [33]. *It can be seen that the DK-GMACC algorithm is a rather broad concept from which many already existing traditional robust adaptive filtering algorithms can be derived.*

Property 6.4 *Recalling* (6.16), *it can be rewritten as*

$$\boldsymbol{w}(m+1) = \boldsymbol{w}(m) + \eta_{\sigma_{+}\sigma_{-},m}|e(m)|^{\gamma-1}\boldsymbol{x}(m), \tag{6.22}$$

when $|e(m)| \to \infty$, *it can be obtained that* $\eta_{\sigma_{+}\sigma_{-},m} \to 0$. *This property demonstrates that a large error will have little effect on the DK-GMACC algorithm, which means that the DK-GMACC algorithm is robust in the face of large outliers or impulsive noise.*

Property 6.5 *Equation* (6.17) *can be seen as a new variable step-size strategy, where the step-size of the algorithm proposed can be varied by using the error signal* [35]. *In the experimental phase, we compared the proposed variable step-size strategy with the fixed step-size of MACC as well as GMACC algorithms and the experimental results are shown in* Fig. 6.3.

Remark 6.2 The computational complexity of the DK-GMACC algorithm is slightly higher than that of the LMP algorithm, and the extra computational effort needed is to calculate the term (6.15). The computational burden is not expensive compared to the performance improvement of the algorithm.

6.5 Theoretical Analysis

The detailed procedure of analyzing the stability and steady-state performance of the DK-GMACC algorithm is shown in this section.

6.5.1 Stability Analysis

To analyze the stability of the DK-GMACC algorithm, the *weight error vector* $\tilde{\boldsymbol{w}}(m)$ need to be defined, that is

$$\tilde{\boldsymbol{w}}(m) = \boldsymbol{w}_{opt} - \boldsymbol{w}(m). \tag{6.23}$$

Combining with (6.16), we have

$$\tilde{\boldsymbol{w}}(m+1) = \tilde{\boldsymbol{w}}(m) - \eta_{\sigma_{+}\sigma_{-}}\phi_{DK-GAC}(e(m))|e(m)|^{\gamma-1}\boldsymbol{x}(m). \tag{6.24}$$

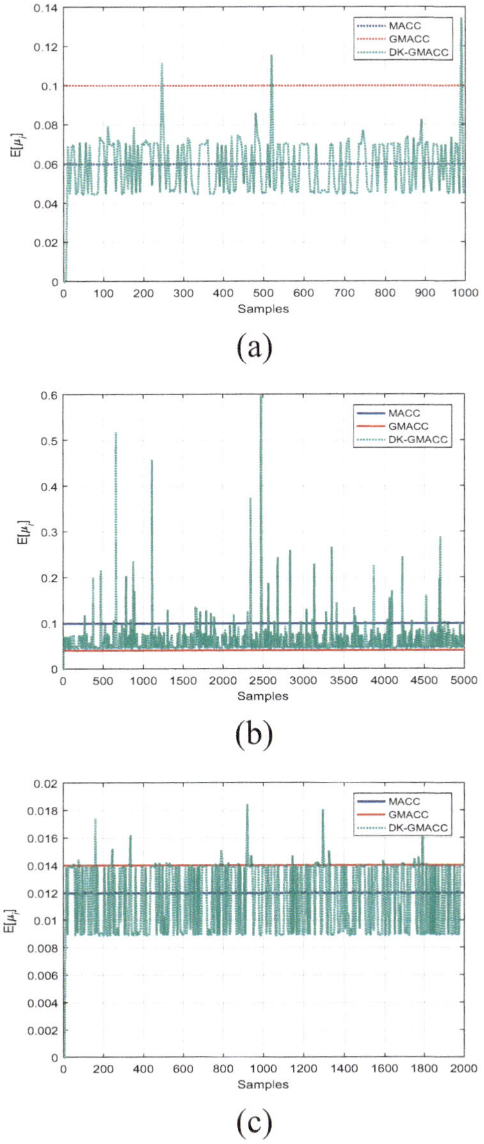

Fig. 6.3 Curves of step-size, **a** the Experiment 1, **b** the Experiment 2, **c** the Experiment 3

After some simple calculations, one can get

$$E\left[\|\tilde{\boldsymbol{w}}(m+1)\|^2\right] = E\left[\|\tilde{\boldsymbol{w}}(m)\|^2\right] - 2\eta_{\sigma_+\sigma_-}E\left[\tilde{\boldsymbol{w}}^T(m)\phi_{DK-GAC}(e(m))|e(m)|^{\gamma-1}\boldsymbol{x}(m)\right] + \eta^2_{\sigma_+\sigma_-}E\left[\left\|\phi_{DK-GAC}(e(m))|e(m)|^{\gamma-1}\boldsymbol{x}(m)\right\|^2\right], \tag{6.25}$$

to ensure that the algorithm is stable, the *weight error vector* needs to satisfy the following relationship,

$$E\left[\|\tilde{\boldsymbol{w}}(m+1)\|^2\right] \le E\left[\|\tilde{\boldsymbol{w}}(m)\|^2\right]. \tag{6.26}$$

Thus, when the step-size μ for all i satisfies the following relation, i.e.,

$$\begin{aligned} &\eta_{\sigma_+\sigma_-}^2 E\left[\left\|\phi_{DK-GAC}(e(m))|e(m)|^{\gamma-1}\boldsymbol{x}(m)\right\|^2\right] \\ &\le 2\eta_{\sigma_+\sigma_-} E\left[\tilde{\boldsymbol{w}}^T(m)\phi_{DK-GAC}(e(m))|e(m)|^{\gamma-1}\boldsymbol{x}(m)\right], \end{aligned} \tag{6.27}$$

Simplifying (6.26) again, we have

$$\eta_{\sigma_+\sigma_-} \le \frac{2E\left[\tilde{\boldsymbol{w}}^T(m)\phi_{DK-GAC}(e(m))|e(m)|^{\gamma-1}\boldsymbol{x}(m)\right]}{E\left[\left\|\phi_{DK-GAC}(e(m))|e(m)|^{\gamma-1}\boldsymbol{x}(m)\right\|^2\right]}. \tag{6.28}$$

At this case, $E[||\tilde{\boldsymbol{w}}(m)||^2]$ is monotonically decreasing and converging, and the adaptive filter finally converges to the steady-state after iterations.

Remark 6.3 From (6.27), it can be seen that the threshold value of the step-size $\eta_{\sigma_+\sigma_-}$ depends on many factors, such as $e(m)$, $v(m)$, σ, α, and etc. Therefore, it is difficult to determine the threshold value before experimentation because the steady-state error is unknown. Usually, a smaller step-size can help the algorithm to converge to the steady-state.

6.5.2 *Steady-State Performance Analysis*

Before proceeding to the steady-state performance analysis of the DK-GMACC algorithm, some assumptions need to be set in advance for the convenience of the analysis [36–42].

Assumption 6.1 $V(m)$ is zero-mean with the variance is σ_v^2, independent, identically, distributed, and is independent of the input $\boldsymbol{x}(m)$.

Assumption 6.2 $E_a(m)$ is zero-mean and independent of the noise $v(m)$.

Assumption 6.3 The filter is long enough, $e_a(m)$ is Gaussian and $||\boldsymbol{x}(m)||^2$ is independent of $\varphi(e(m)) = \phi_{DK-GAC}(e(m))|e(m)|^{\gamma-1}$.

Define the a priori *error* and the a posteriori *error* as

$$e_a(m) = \tilde{\boldsymbol{w}}^T(m)\boldsymbol{x}(m), \tag{6.29}$$

$$e_p(m) = \tilde{\boldsymbol{w}}^T(m+1)\boldsymbol{x}(m). \tag{6.30}$$

Combining $d(m) = \boldsymbol{w}_{opt}^T\boldsymbol{x}(m) + v(m)$ and (6.28), e_i can be rewritten as

$$e_i = \tilde{\mathbf{w}}_i^{\mathrm{T}}\boldsymbol{n}_i + v_i = e_{a,i} + v_i. \tag{6.31}$$

Define the steady-state mean square error (MSE) and the excess mean square error (EMSE) as

$$E_{\mathrm{MSE}} = \lim_{i\to\infty} E\big[e^2(m)\big], \tag{6.32}$$

$$E_{\mathrm{EMSE}} = \lim_{i\to\infty} E\big[e_a^2(m)\big]. \tag{6.33}$$

Using the relation in (6.31) and combining (6.32) and (6.33), there are

$$E_{\mathrm{MSE}} = E_{\mathrm{EMSE}} + \sigma_v^2. \tag{6.34}$$

Remark 6.4 In the steady-state performance analysis of adaptive filtering, EMSE can filter out the direct effect of measurement noise on the steady-state performance and is therefore widely used as a measure of steady-state performance.

When $m \to \infty$, i.e., the filter reaches the steady-state, $E[||\tilde{\boldsymbol{w}}(m+1)||^2] = E[||\tilde{\boldsymbol{w}}(m)||^2]$. Combining (6.16) and (6.23), this gives

$$\begin{aligned}&\eta_{\sigma_+\sigma_-}^2 E\Big[\big\|\phi_{\mathrm{DK-GAC}}(e(m))|e(m)|^{\gamma-1}\boldsymbol{x}(m)\big\|^2\Big]\\ &\quad= 2\eta_{\sigma_+\sigma_-}E\big[\tilde{\boldsymbol{w}}^T(m)\phi_{\mathrm{DK-GAC}}(e(m))|e(m)|^{\gamma-1}\boldsymbol{x}(m)\big].\end{aligned} \tag{6.35}$$

Recalling $\varphi(e(m))$, (6.34) can be rewritten as

$$\eta_{\sigma_+\sigma_-}^2 E\big[\|\varphi(e(m))\boldsymbol{x}(m)\|^2\big] = 2\eta_{\sigma_+\sigma_-}E[\varphi(e(m))e_a(m)]. \tag{6.36}$$

Remark 6.5 The distributions of $e_a(m)$ and $e_p(m)$ in the steady-state are independent of time i. The time-indexed subscripts are omitted in the next derivation.

By employing the Assumption 3, one can simplify (6.36) as

$$\eta_{\sigma_+\sigma_-}^2 \mathrm{Tr}(\boldsymbol{R}_{xx})E\big[\|\varphi(e)\|^2\big] = 2\eta_{\sigma_+\sigma_-}E[\varphi(e)e_a]. \tag{6.37}$$

Applying Assumptions 6.1, 6.2, taking the Taylor expansion of $\varphi(e)$ with respect to e_a around v, that is

$$\varphi(e) = \varphi(e_a + v) = \varphi(v) + \varphi'(v)e_a + \frac{1}{2}\varphi''(v)e_a^2 + o\big(e_a^2\big), \tag{6.38}$$

where $\varphi'(v)$ and $\varphi''(v)$ are the first and second derivatives, and $o(e_a^2)$ denotes the third and higher-order terms.

By the nature of the expectation operator and with $o(e_a^2)$ sufficiently small, it can be further derived

$$E[\varphi(e)e_a] = E\left[\varphi(v)e_a + \varphi'(v)e_a^2 + o\left(e_a^2\right)\right] \approx E\left[\varphi'(v)\right]E_{\mathrm{EMSE}}, \tag{6.39}$$

$$E\left[\varphi^2(e)\right] \approx E\left[\varphi^2(v)\right] + E\left[\varphi(v)\varphi''(v) + \left|\varphi'(v)\right|^2\right]E_{\mathrm{EMSE}}, \tag{6.40}$$

Substitute (6.39), (6.40) into (6.37), one can obtain

$$E_{\mathrm{EMSE}} = \frac{\eta_{\sigma_+\sigma_-}(v)\mathrm{Tr}(\boldsymbol{R}_{xx})E\left[\varphi^2(v)\right]}{2E[\varphi'(v)] - \eta_{\sigma_+\sigma_-}(v)\mathrm{Tr}(\boldsymbol{R}_{xx})E\left[\varphi(v)\varphi''(v) + |\varphi'(v)|^2\right]}, \tag{6.41}$$

where $\eta_{\sigma_+\sigma_-}(v) = \begin{cases} \mu(\alpha_1\beta_{1,1} + \dfrac{\alpha_2\beta_{2,1}}{|v|}), & \text{if } v \geq 0 \\ \mu(\alpha_1\beta_{1,2} + \dfrac{\alpha_2\beta_{2,2}}{|v|}), & \text{if } v < 0 \end{cases}$, $\mathrm{Tr}(\cdot)$ stands for the trace of the matrix, $\boldsymbol{R}_{xx} = [\boldsymbol{x}(m)\boldsymbol{x}^T(m)]$. With some simple math calculations, $\varphi(v)$, $\varphi'(v)$, $\varphi''(v)$, $\varphi^2(v)$ in (6.41) can be computed as

$$\varphi(v) = \begin{cases} f_1(v)|v|^{\gamma-1}, \text{ if } v \geq 0 \\ f_2(v)|v|^{\gamma-1}, \text{ if } v < 0 \end{cases}, \tag{6.42}$$

where $f_1(v) = \exp(-(\alpha_1\beta_{1,1}|v|^{\gamma} + \alpha_2\beta_{2,1}|v|^{\gamma-1}))$, $f_2(v) = \exp(-(\alpha_1\beta_{1,2}|v|^{\gamma} + \alpha_2\beta_{2,2}|v|^{\gamma-1}))$.

$$\varphi'(v) = \begin{cases} \omega_1(v)f_1(v), \text{ if } v \geq 0 \\ \omega_2(v)f_2(v), \text{ if } v < 0 \end{cases}, \tag{6.43}$$

where

$$\omega_1(v) = -\alpha_1\beta_{1,1}\gamma v^{\gamma-1} + \left(1 - \alpha_2\beta_{2,1}\right)(\gamma - 1)v^{\gamma-2}, \tag{6.44}$$

$$\omega_2(v) = -\alpha_1\beta_{1,2}\gamma v^{\gamma-1} + \left(1 - \alpha_2\beta_{2,2}\right)(\gamma - 1)v^{\gamma-2}, \tag{6.45}$$

$$\varphi''(v) = \begin{cases} \omega_3(v)f_1(v) + \omega_4(v)f_1(v), \text{ if } v \geq 0 \\ \omega_5(v)f_2(v) + \omega_6(v)f_2(v), \text{ if } v < 0 \end{cases}, \tag{6.46}$$

$$\omega_3(v) = -\alpha_1\beta_{1,1}\gamma(\gamma - 1)v^{\gamma-2} + \left(1 - \alpha_2\beta_{2,1}\right)(\gamma - 1)(\gamma - 2)v^{\gamma-3}, \tag{6.47}$$

$$\omega_4(v) = -\left(\alpha_1\beta_{1,1}\gamma v^{\gamma-1} + \alpha_2\beta_{2,1}(\gamma-1)v^{\gamma-2}\right), \tag{6.48}$$

$$\omega_5(v) = -\alpha_1\beta_{1,2}\gamma(\gamma-1)v^{\gamma-2} + \left(1-\alpha_2\beta_{2,2}\right)(\gamma-1)(\gamma-2)v^{\gamma-3}, \tag{6.49}$$

$$\omega_6(v) = -\left(\alpha_1\beta_{1,2}\gamma v^{\gamma-1} + \alpha_2\beta_{2,2}(\gamma-1)v^{\gamma-2}\right), \tag{6.50}$$

$$\varphi^2(v) = \begin{cases} f_1^2(v)|v|^{2\gamma-2}, \text{ if } v \geq 0 \\ f_2^2(v)|v|^{2\gamma-2}, \text{ if } v < 0 \end{cases}, \tag{6.51}$$

According to the above equations, the closed-form solution of EMSE can be obtained (Fig. 6.4).

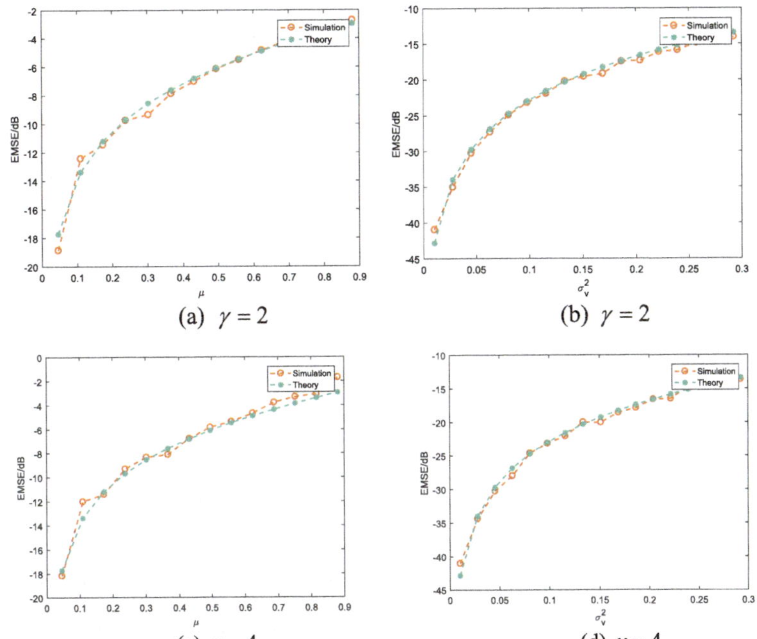

Fig. 6.4 Results of the theoretical analysis values compared to the experimental simulation values ($\sigma_v^2 = 0.8$ in (**a**) and (**c**). $\mu = 0.5$ in (**b**) and (**d**))

6.6 Computer Simulation Examples

In this section, we validate the effectiveness of the methods used for the theoretical analysis of the DK-GMACC algorithm. The performance of the proposed algorithm is also tested in the context of system identification, AEC and wind prediction. The parameter settings of all the algorithms in the experiment uphold the principle of ensuring that all have the same initial convergence speed. All learning curves are obtained by averaging over 100 independent runs.

6.6.1 Theoretical Verification

For the theoretical experimental validation, the experimental environment is set up as follows: the unknown weight vector $\boldsymbol{w}_{opt}$ is a white Gaussian with zero-mean and variance 1.0, the measurement noise is assumed to a zero-mean Gaussian noise sequence, and the input signal is a zero-mean Gaussian process with variance 1.0. The parameters are set to $\alpha_1 = 0.998$, $\sigma_{1,+} = 3$, $\sigma_{1,-} = 1.2$, $\sigma_{2,+} = 2.4$, and $\sigma_{2,-} = 1.2$. The length of the filter is 8. The EMSE is calculated by (6.41). We carry out experiments to verify the validity of the theoretical analytical approach of the algorithm under $\gamma = 2$ and $\gamma = 4$, respectively. In addition, we conduct experimental comparisons from the perspective of different values of step-sizes and noise variance, respectively.

6.6.2 System Identification

In order to test the performance of the proposed algorithm, we conduct experimental simulations on the convergence speed and steady-state performance of the algorithm in three different environments.

Experiment 1: $\boldsymbol{w}_{opt}$ and the input signal are consistent with the experimental setting of the previous section. The measurement noise is assumed to a zero-mean Gaussian noise sequence with variance 0.8.

Experiment 2: $\boldsymbol{w}_{opt}$ and the input signal are consistent with the experimental setting of the Experiment 1. The measurement noise is generated by the $F(5, 14)$, where $F(n, m)$ denotes the F-distribution with degrees of freedom n and m.

Experiment 3: The weight vector of the unknown system is generated by Gaussian distribution with a mean 0.3 and variance 0.1. The input signal is a zero-mean Gaussian process with variance 1.0. The measurement noise is generated by an asymmetric noise model $v(m) = (1 - \theta(m))P(m) + \theta(m)Q(m)$, where $\theta(m)$ is a binary process with $\Pr\{\theta(m) = 1\} = p$, $\Pr\{\theta(m) = 0\} = 1 - p$, and $0 \le p \le 1$ are the probability of occurrence. $P(m) = Z_1(m) + Z_2(m)$, $Z_1(m)$, $Z_2(m)$, and $Q(m)$ are white Gaussian noises with zero-mean and variance 0.8, 8, and 400 (very large variances represent large outliers), respectively. The normalized mean-square-deviation (NMSD) is used to test the behaviors of the algorithms, where NMSD denotes as: $\text{NMSD} = 10\log_{10}(||\boldsymbol{w}_{opt} - \boldsymbol{w}||_2^2 / ||\boldsymbol{w}||_2^2)$.

The results in Figs. 6.5 and 6.6 show that the proposed algorithm possesses better performance compared to MCC, MMKCC, MMKACC, MACC, GMACC algorithms (where MMKCC was proposed in [22] and MMKACC is obtained by applying MMKCC to MACC). The MCC algorithm performs weakly as compared to several other algorithms. The proposed algorithm performs optimally both in terms of convergence speed, steady-state and tracking performance.

In order to avoid the influence of step-size and the measurement noise variance on the performance comparison, we have again compared the steady-state performance of the algorithms under different values of step-size and noise variance. Figure 6.7 shows the performance comparison results under different step size conditions. Figure 6.8 shows the performance comparison results under different noise variance conditions, where the experimental conditions in Fig. 6.8c, d are set as follows: Fig. 6.8c the variance of $Z_2(m)$ is fixed at 8, the variance of $Z_1(m)$ is time-varying, Fig. 6.8d the variance of $Z_1(m)$ is fixed at 0.8, the variance of $Z_1(m)$ is

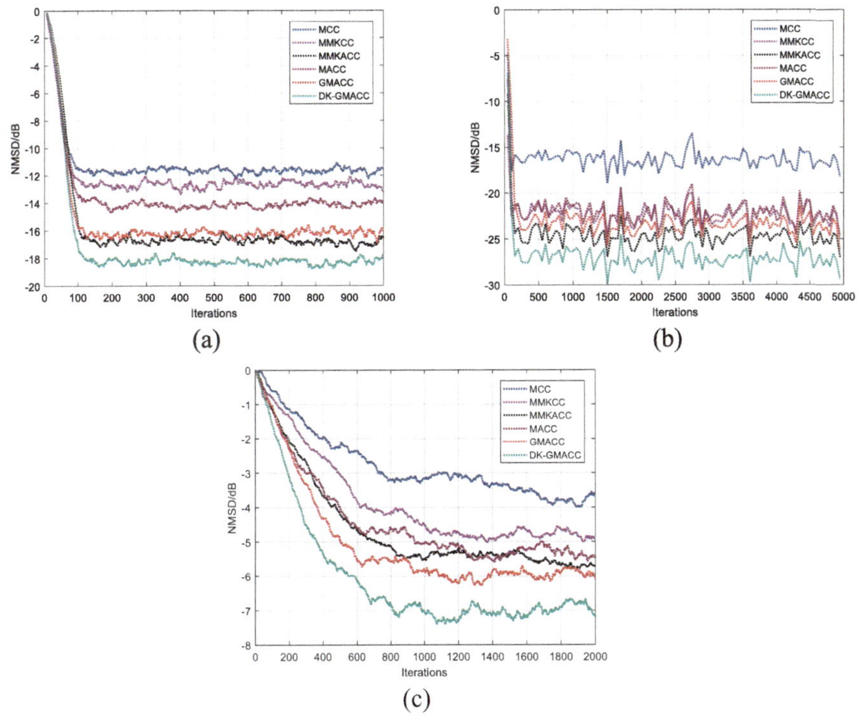

Fig. 6.5 Performance curves of different algorithms, **a** Experiment 1, **b** Experiment 2, **c** Experiment 3

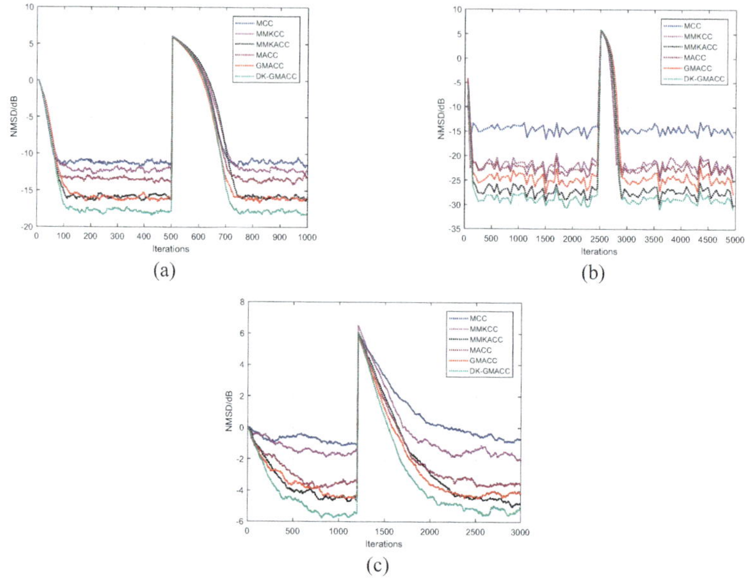

Fig. 6.6 Tracking performance curves of different algorithms, **a** Experiment 1, **b** Experiment 2, **c** Experiment 3

time-varying. The results in Figs. 6.7 and 6.8 show that the steady-state performance of the algorithms diminishes as the step-size and noise variance increase. However, the proposed algorithm still maintains the best steady-state performance for different values of step-size and noise variance.

6.6.3 Echo Cancellation Scenario

In the echo cancellation, the performance of DK-GMACC and other algorithms is evaluated when the algorithms are used in single-talk and double-talk scenarios. The NMSD and echo return loss enhancement (ERLE) are used to evaluate the performance of the algorithms, where ERLE is defined as [43, 44], $y(m)$ is the true echo and $\tilde{e}(m)$ denotes the error between $y(m)$ and the output of the filter. In addition, the tracking performance of the algorithms is tested at the same time. To simulate changes in the echo paths and to evaluate the tracking ability of the algorithm, the echo path impulse responses were shifted to the right by 12 bits at intermediate iterations. In this experiment, the filter length is 256, and the impulse response of the echo path is shown in Fig. 6.9. The measurement noise consists of the superposition of two noises, i.e., $v(m) = a(m) + b(m)$, where $a(m)$ is a zero-mean Gaussian noise sequence with variance 0.01, $b(m)$ is generated by the $F(5, 14)$.

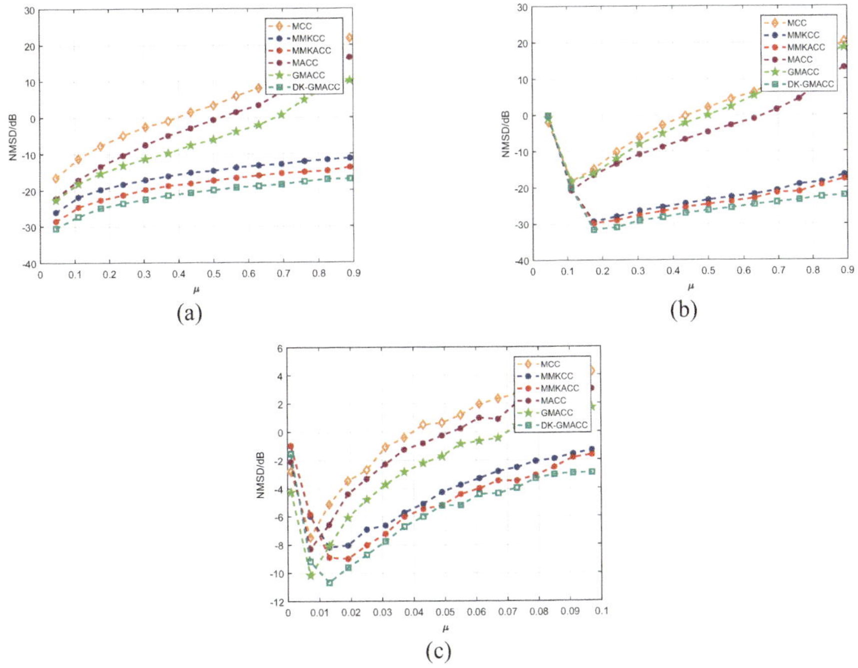

Fig. 6.7 Steady-state performance of different algorithms with different step-size, **a** Experiment 1, **b** Experiment 2, **c** Experiment 3

Figure 6.10 illustrates the NMSD and ERLE performance curves of the proposed algorithm and other competing algorithms for mono and two-channel scenarios. It can be seen that MCC has the worst performance. Compared with the existing robustness algorithms, the proposed DK-GMACC algorithm has the lowest NMSD and highest ERLE values. In other words, it has the best echo cancellation performance. The DK-GMACC algorithm maintains its performance advantage in both single-talk and dual-talk scenarios.

6.6.4 Wind Prediction

In this subsection, the performance of the DK-GMACC and other algorithms is tested in the wind prediction scenario. The sampled set was recorded at the Institute of Industrial Science, University of Tokyo [45] and sampled at 50 Hz. The high wind-speed region (north–south direction) is selected as the data set.

The magnitude of wind signal is shown in Fig. 6.11. The first 3000 samples are used as the training set, and the last 2000 samples are used as the testing set. In order to measure the performance of the algorithm, the prediction gain R_p is employed [38], which is defined as

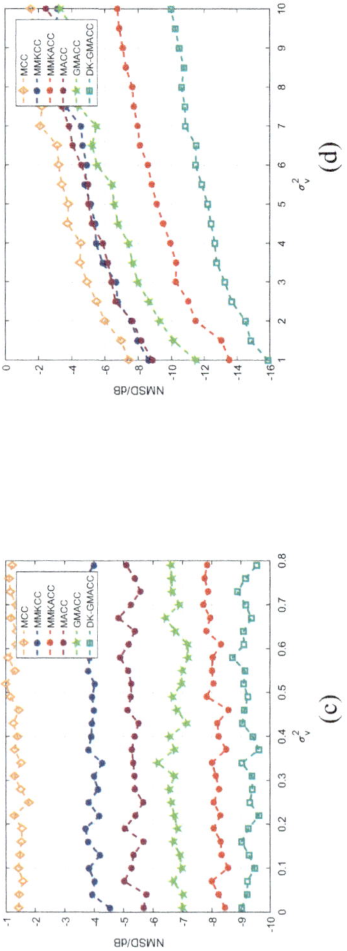

Fig. 6.8 Steady-state performance of different algorithms with different noise variances, **a** Experiment 1, **b** Experiment 2, **c** Experiment 3, **d** Experiment 3

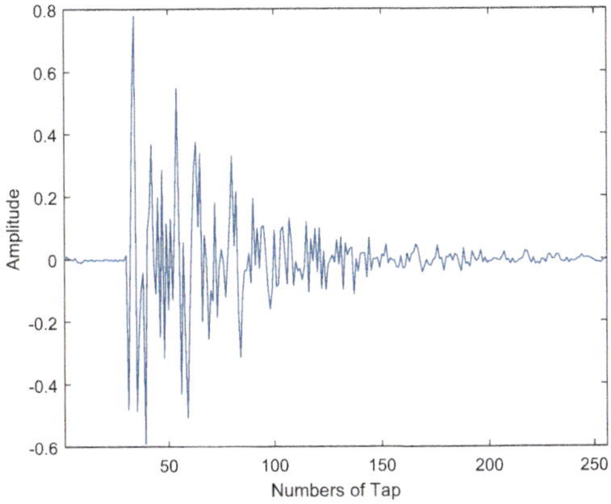

Fig. 6.9 Impulse response of the echo path

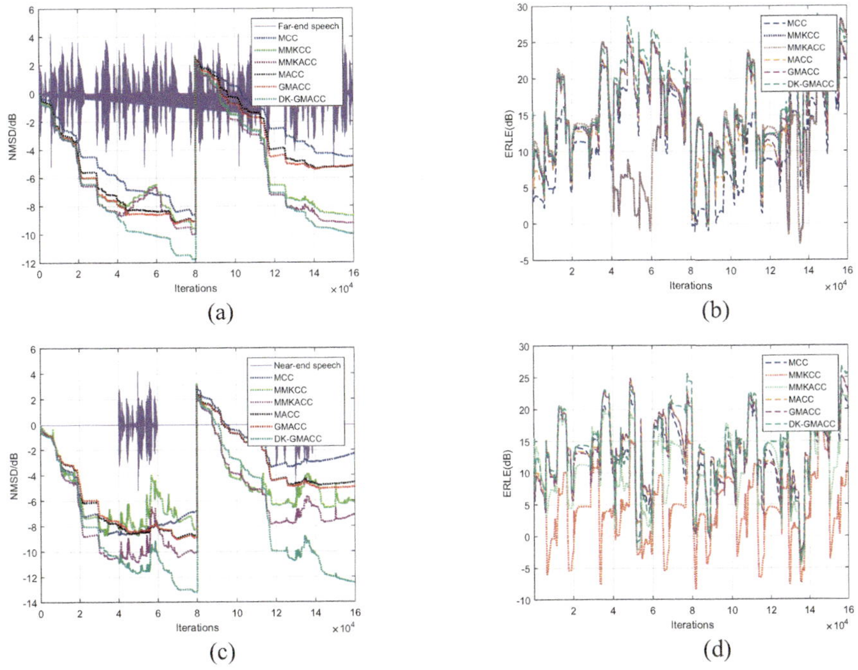

Fig. 6.10 AEC performance of DK-GMACC and other algorithms in single-talk and double-talk scenarios. Single-talk scenario ((**a**), (**b**)). Double-talk scenario ((**c**), (**d**))

$$R_p = 10\log 10\left(\frac{\sigma_{v_n}^2}{\tilde{\sigma}_e^2}\right), \tag{6.52}$$

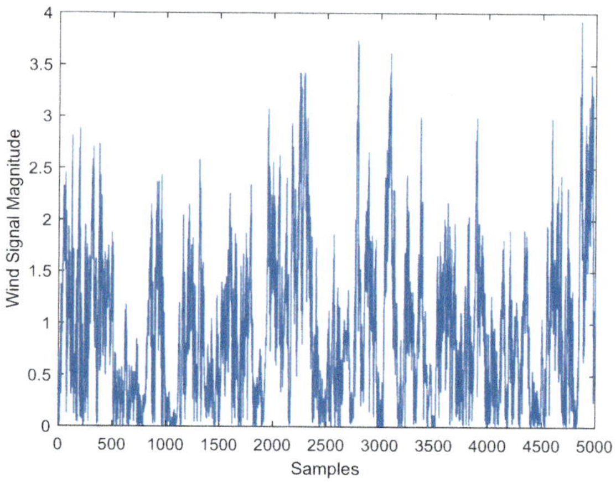

Fig. 6.11 Wind signal magnitude

where $\sigma_{v_n}^2$ is the variance of the input signal $\{v_n(m)\}$, and $\tilde{\sigma}_e^2$ denotes the estimated variance of $\{e(m)\}$.Then the window size W_s of the moving average filter varied according to [45], i.e., $W_s = [1\,2\,10\,20\,60\,300]$.

Figure 6.12 shows the prediction gain comparison of the DK-GMACC and other algorithms with the different values of W_s. And Fig. 6.12 illustrates that the proposed algorithm has the prediction gain better than other existing comparison algorithm.

Finally, we conducted an experimental test on the prediction ability of six algorithms on the wind signal, and the window size was chosen to be 60. The results are shown in Fig. 6.13. From the Fig. 6.13, it can be seen that the prediction results of the other several algorithms have some deviation compared to the DK-GMACC algorithm. The proposed algorithm has the best match with the input signal.

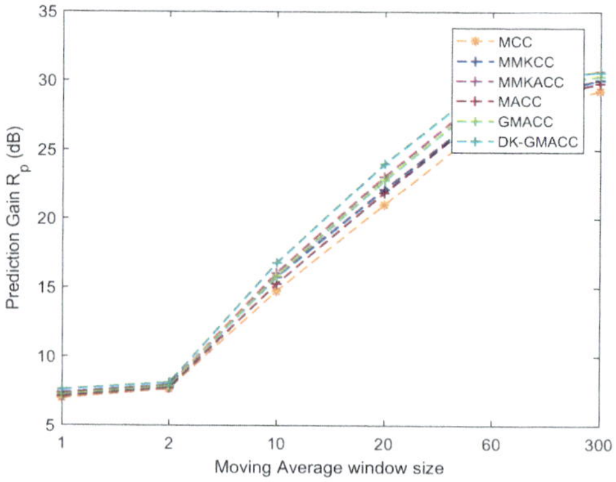

Fig. 6.12 Prediction gains of DK-GMACC and other algorithms

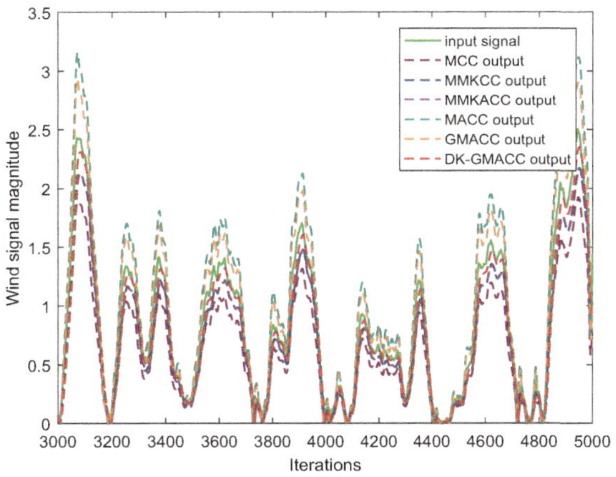

Fig. 6.13 Prediction comparison of DK-GMACC and other algorithms

6.7 Summary

In this chapter, a dual-kernel generalized maximum asymmetric correntropy adaptive algorithm is proposed. By adjusting the occupancy ratio of two different kernel functions, the proposed dual-kernel maximum correntropy can be adapted to different noise environments, which in turn can improve the performance of the symmetric correntropy algorithms in the face of different non-Gaussian environments, as well as asymmetric signals or noise. A generalized dual-kernel form of correntropy also improves the generalization as well as applicability of the algorithm. Secondly, in the algorithm proposal work, we give a new variable step size strategy using error signal. Subsequently, we theoretically analyze the stability as well as steady-state performance of the proposed algorithm. In the experimental simulation section, we test the performance of DK-GMACC algorithm in different scenarios, i.e., system identification, acoustic echo cancellation, wind prediction, and different noise environments. The simulation results show that the proposed algorithm has the best performance under different experimental conditions. Finally, our proposed dual-kernel correntropy will provide a new idea for future algorithms in complex-valued domain, geometric-algebraic domain, and affine-projection algorithm, which will also lay a foundation for us to carry out related research in the future.

References

1. M. Godavarti, A.O. Hero, Partial update LMS algorithms. IEEE Trans. Signal Process. **53**(7), 2382–2399 (2005)
2. J. Benesty, H. Rey, L.R. Vega, S. Tressens, A nonparametric VSS NLMS algorithm. IEEE Signal Process. Lett. **13**(10), 581–584 (2006)
3. D. Mandic, V.S.L. Goh, *Complex Valued Nonlinear Adaptive Filters: Non-Circularity, Widely Linear and Neural Models* (Wiley, New York, NY, USA, 2009)

4. X. Long, H. Zhao, X. Hou, W, Quan, A variable regularization parameter widely linear complex-valued NLMS algorithm: Performance analysis and wind prediction. Signal Process. **201** (2022)
5. S. Lv, H. Zhao, X, He, Geometric algebra based least mean m-estimate robust adaptive filtering algorithm and its transient performance analysis. Signal Process. **189** (2021)
6. X. Long, H. Zhao, X. Hou, Hybrid geometric algebra correntropy: Definition and application to robust adaptive filtering. IEEE Trans. Circuits Syst. II Exp. Briefs **71**(2), 952–956 (2024)
7. C. Cheong Took, D.P. Mandic, A quaternion widely linear adaptive filter. IEEE Trans. Signal Process. **58**(8), 4427–4431 (2010)
8. M. Dahl, I. Claesson, Acoustic noise and echo cancelling with microphone array. IEEE Trans. Veh. Technol. **48**(5), 1518–1526 (1999)
9. C.-H. Yang, J.-C. Wang, J.-F. Wang, C.-H. Wu, K.-H. Chang, Design and implementation of subspace-based speech enhancement under in-car noisy environments. IEEE Trans. Veh. Technol. **57**(3), 1467–1479 (2008)
10. J.G. Ryan, R.A. Goubran, Application of near-field optimum microphone arrays to hands-free mobile telephony. IEEE Trans. Veh. Technol. **52**(2), 390–400 (2003)
11. S. Liu, T. Zhang, L. Zhang, A SINS aided correct method for USBL range based on maximum correntropy criterion adaptive filter. IEEE Trans. Instrum. Meas. **71**, 1–13 (2022)
12. K. Zhang, H. Wang, H. Zhang, N. Luo, J. Ren, Target tracking of UUV based on maximum correntropy high-order UGHF. IEEE Trans. Instrum. Meas. **72**, 1–16 (2023)
13. W. Ma, Q. Yan, P. Guo, B. Chen, Robust state estimation of integrated electric-heat system based on physics-guided deep learning with maximum correntropy criterion. IEEE Trans. Instrum. Meas. **74**, 1–12 (2025)
14. B. Chen, L. Xing, H. Zhao, N. Zheng, J.C. Principe, Generalized correntropy for robust adaptive filtering. IEEE Trans. Signal Process. **64**(13), 3377–3387 (2016)
15. J.C. Principe, *Information Theoretic Learning: Renyi's Entropy and Kernel Perspectives*. Springer Science & Business Media, Berlin, Germany, 2010)
16. X. Long, H. Zhao, X. Hou, A novel combinatoric correntropy algorithm: Properties and its performance analysis. IEEE Trans. Circuits Syst. II Exp. Briefs **69**(12), 5184–5188 (2022)
17. W. Liu, P.P. Pokharel, J.C. Principe, Correntropy: properties and applications in non-Gaussian signal processing. IEEE Trans. Signal Process. **55**(11), 5286–5298 (2007)
18. L. Shi, Y. Lin, Convex combination of adaptive filters under the maximum correntropy criterion in impulsive interference. IEEE Signal Process. Lett. **21**(11), 1386–1388 (2014)
19. B. Chen, L. Xing, H. Zhao, S. Du, J.C. Príncipe, Effects of outliers on the maximum correntropy estimation: a robustness analysis. IEEE Trans. Syst., Man, Cybern. Syst. **51**(6), 4007–4012 (2021)
20. Y. Zheng, B. Chen, S. Wang, W. Wang, Broad learning system based on maximum correntropy criterion. IEEE Trans. Neural Netw. Learn. Syst. **32**(7), 3083–3097 (2021)
21. Y. He, F. Wang, Y. Li, J. Qin, B. Chen, Robust matrix completion via maximum correntropy criterion and half-quadratic optimization. IEEE Trans. Signal Process. **68**, 181–195 (2020)
22. B. Chen, Y. Xie, X. Wang, Z. Yuan, P. Ren, J. Qin, Multikernel correntropy for robust learning. IEEE Trans. Cybern. **52**(12), 13500–13511 (2022)
23. H. Sun, X. Yang, H. Gao, A spatially constrained shifted asymmetric Laplace mixture model for the grayscale image segmentation. Neurocomputing **331**, 50–57 (2019)
24. W. You, Y. Guo, H. Zhu, Y. Tang, Oil price shocks, economic policy uncertainty and industry stock returns in China: asymmetric effects with quantile regression. Energy Econ. **68**, 1–18 (2017)
25. X. Zhao, S. Fu, Y. Tian, K. Zhao, Asymmetric and robust loss function driven least squares support vector machine. Knowl.-Based Syst. **258** (2022)
26. B. Chen, Y. Xie, Z. Li, Y. Li, P. Ren, Asymmetric correntropy for robust adaptive filtering. IEEE Trans. Circuits Syst. II Exp. Briefs **69**(3), 1922–1926 (2022)
27. Q. Hua, M. Wang, J. Zhao, S. Zhao, T. Li, P. Yue, Generalized asymmetric correntropy for robust adaptive filtering: a theoretical and simulation study. Remote Sens. **14**(15) (2022)
28. T. Elguebaly, N. Bouguila, Finite asymmetric generalized Gaussian mixture models learning for infrared object detection. Comput. Vis. Image Underst. **117**(12), 1659–1671 (2013)

29. Y. Peng, X. Huang, C. Yu, Q. Mao, Y. Li, Affine-projection maximum asymmetric correntropy criterion algorithm under system identifications. In: 2023 IEEE 11th International Conference on Computer Science and Network Technology (ICCSNT), Dalian, China, pp. 527–530 (2023)
30. H. Yin, J. Mei, F. Dong,·G. Qian,·S. Wang, Asymmetric complex correntropy for robust adaptive filtering. Circ. Syst. Signal Process **41**, 4692–4706 (2022)
31. S. Han, S. Rao, D. Erdogmus, K.H. Jeong, J. Principe, A minimum-error entropy criterion with self-adjusting step-size (MEE-SAS). Signal Process. **87**(11) (2007)
32. X. Hou, H. Zhao, X. Long, H. C. So, "Computationally efficient robust adaptive filtering algorithm based on improved minimum error entropy criterion with fiducial points," *ISA Transactions*, ISSN: 0019–0578,2024.
33. S. Haykin, *Adaptive Filter Theory*, 5th edn. (Pearson, London, U.K., 2014)
34. S.-C. Pei, C.-C. Tseng, Least mean p-power error criterion for adaptive fir filter. IEEE J. Sel. Areas Commun. **12**(9), 1540–1547 (1994)
35. S. Radhika, A. Chandrasekar, Maximum correntropy criteria adaptive filter with adaptive step size. In: 2018 IEEE International Conference on Computational Intelligence and Computing Research (ICCIC), Madurai, India, pp. 1–4 (2018)
36. Y. Zhu, H. Zhao, X. Zeng, B. Chen, Robust generalized maximum correntropy criterion algorithms for active noise control. IEEE/ACM Trans. Audio, Speech, Lang. Process. **28**, 1282–1292 (2020)
37. W. Shi, Y. Li, Y. Wang, Noise-free maximum correntropy criterion algorithm in non-gaussian environment. IEEE Trans. Circuits Syst. II, Exp. Briefs **67**(10), 2224–2228 (2020)
38. H. Zhao, D. Liu, S. Lv, Robust maximum correntropy criterion subband adaptive filter algorithm for impulsive noise and noisy input. IEEE Trans. Circuits Syst. II, Exp. Briefs **69**(2), 604–608 (2022)
39. W. Wang, H. Zhao, X. Zeng, K. Doğançay, Steady-state performance analysis of nonlinear spline adaptive filter under maximum correntropy criterion. IEEE Trans. Circuits Syst. II, Exp. Briefs **67**(6), 1154–1158 (2020)
40. B. Chen, L. Xing, J. Liang, N. Zheng, J.C. Príncipe, Steady-state mean-square error analysis for adaptive filtering under the maximum correntropy criterion. IEEE Signal Process. Lett. **21**(7), 880–884 (2014)
41. H. Zhao, Y. Gao, Y. Zhu, Robust subband adaptive filter algorithms-based mixture correntropy and application to acoustic echo cancellation. IEEE/ACM Trans. Audio, Speech, Lang. Process. **31**, 1223–1233 (2023)
42. Z. Zheng, Z. Liu, J. Zhao, Bias-compensated sparsity-aware NLMM algorithms for robust adaptive echo cancellation. IEEE Trans. Circuits Syst. I Regul. Pap. **67**(7), 2383–2396 (2020)
43. J. Benesty, C. Paleologu, T. Gansler, S. Ciochina, A perspective on stereophonic acoustic echo cancellation, in *Springer Topics in Signal Process.*, vol. 4. (Springer, Berlin, Germany, 2011), pp. 9–11
44. F. Huang, J. Zhang, S. Zhang, Affine projection versoria algorithm for robust adaptive echo cancellation in hands-free voice communications. IEEE Trans. Veh. Technol. **67**(12), 11924–11935 (2018)
45. D. Mandic, S. Javidi, S.L. Goh, A. Kuh, K. Aihara, Complex-valued prediction of wind profile using augmented complex statistics. Renewable Energy **34**(1), 196–201 (2009)

Chapter 7
Hybrid Geometric Algebra Correntropy Criterion

7.1 Introduction

In linear algebraic space, adaptive filters are usually formulated by standard vector calculus and the adaptive filtering algorithms (AFAs) are designed based on the underlying mean square error criterion (MSE), in which the least mean square (LMS) type algorithms have become one of the most popular research hotspots due to their simplicity of structure and ease of implementation [1, 2]. However, they can be severely degraded when the system is interfaced by non-Gaussian noise disturbances. Therefore, it is very crucial to study robust adaptive filtering algorithms against non-Gaussian noise interferences; for example, on the basis of information learning theory, the concept of entropy is proposed. Since entropy contains information about the higher-order moments of the signal, the adaptive filtering algorithms based on the maximum correntropy criterion and the minimum error entropy criterion are both robust in the face of non-Gaussian disturbances [3–10].

In practice, many applications involving engineering, including multi-dimensional signal processing and machine learning, are defined in the geometric algebra (GA) domain. The geometric product, as a significant operation, is a product that can map vectors onto polytopes, scalars, and hypersurfaces [11]. GA can be used to avoid information loss and reserve the correlations among multiple dimensions. However, the classical linear AFAs no longer apply in these areas. In recent years, with the rapid development of signal processing techniques, AFAs have been developed that can perform image processing as well as multi-dimensional signal processing. For example, using the corresponding properties of GA, GA-LMS algorithms based on MSE have been proposed and applied in different areas such as parameter estimation of multi-dimensional vector values and wind speed prediction [11–14]. Specifically, the landmark GA-based LMS algorithm was devised and applied to 3D point cloud registration [11]. The performance of the GA-LMS was analyzed [13]. The modified Huber function was extended to the GA space and the GA-LMM algorithm was derived by using the instantaneous gradient-descent

H. Zhao et al., *Information Theoretic Learning-Based Filter*,
https://doi.org/10.1007/978-3-032-29623-8_7

method, which demonstrates excellent robustness in the face of non-Gaussian noise interference [15]. However, it still suffers from high steady-state error under parameter estimation circumstances. Thus, to improve the shortcomings of the GA-LMS and GA-LMM algorithms, by employing geometric algebra correntropy (GAC), the robust adaptive filtering algorithm (MGACC) is proposed in [16].

To this end, to further obtain performance improvement under non-Gaussian noise interference for parameter estimation, firstly, we define the HGAC through the lens of probability density function and use the HGAC to derive an adaptive algorithm for the estimation of system parameters with multi-dimensional vector values. In addition, to increase the flexibility of the GAC, we use an HGAC. Furthermore, the step size of the proposed algorithm is also variable due to the introduction of the proportionality parameter. Finally, simulations with two different non-Gaussian noise models also demonstrate the better convergence and steady-state performance of the proposed algorithm compared to other competitive algorithms.

7.2 Hybrid Geometric Algebra Correntropy Criterion

Assuming that the GA $\boldsymbol{G}_{m,n}$ is an algebra over the vector space $\mathbb{V}_S^{a+b}$ with the dimension $\mathrm{Dim} = 2^{a+b}$ and constituted by the implementations of the geometric product over the vector basis. Then we define the geometric product of vectors m and n as $mn \triangleq m \cdot n + m \wedge n,\ m, n \subset \mathbb{V}_S^{a+b}$, where $(\cdot)$ denotes the inner product and $(\wedge)$ denotes the outer product. The inner product of vectors m and n is defined as

$$m \cdot n \triangleq |m||n|cos\vartheta, \tag{7.1}$$

where $|\cdot|$ denotes the magnitude of a vector, ϑ is the angle between m and n. The outer product is also defined as

$$m \wedge n \triangleq I_{m,n}|m||n|sin\vartheta, \tag{7.2}$$

where $I_{m,n}$ is the unit bivector that denotes the orientation of the hyperplane $m \wedge n$. It is worth noting that since the inner product is a scalar, the inner product is commutative, i.e., $m \cdot n = n \cdot m$; but the outer product is non-commutative because $m \wedge n = -n \wedge m$. Thus, the geometric product of vectors m and n is usually non-commutative.

Additionally, $\boldsymbol{G}_{m,n}$ can be constructed by the application of the geometric product of the canonical orthonormal basis COB. considering have an ordered set $(\varepsilon_1, \ldots, \varepsilon_a, \varepsilon_{a+1}, \ldots, \varepsilon_{a+b})$ that is composed of the COB, in which $\varepsilon_i\varepsilon_j$ need satisfies

$$\varepsilon_i\varepsilon_j = \begin{cases} \varepsilon_{ij} & \text{for } i \neq j \\ 1 & \text{for } i{=}j\ (i,j \subset [1, 2, \ldots, a]) \\ -1 & \text{for } i{=}j\ (i,j \subset [a+1, \ldots, a+b]) \end{cases} \quad . \tag{7.3}$$

Thus, the basis for $\boldsymbol{G}_{m,n}$ can be expressed as 1, $\{\varepsilon_i\}$, $\{\varepsilon_i \wedge \varepsilon_j\}$, $\{\varepsilon_i \wedge \varepsilon_j \wedge \varepsilon_k\}$, …, $\{\varepsilon_1 \wedge \varepsilon_2 \wedge, \ldots, \wedge \varepsilon_{a+b}\}$, where $i, j, k = 1, 2, \ldots, N$ and $i < j < k$, and the dimension of the $\boldsymbol{G}_{m,n}$ is $2^{(a+b)}$. In the following parts, we only consider the case that the Clifford geometric algebra of three-dimension Euclidean vector space, i.e., $a = 3$, $b = 0$. Meanwhile, the above result of this chapter can be easily extended to the case when Dim ≥ 4. Then, the geometric algebra with three dimension is denoted as $\boldsymbol{G}(\mathbb{R}^3)$. The basis $\{\varepsilon_1, \varepsilon_2, \varepsilon_3\}$ of $\mathbb{R}^3$ generates the basis $\{1, \varepsilon_1, \varepsilon_2, \varepsilon_3, \varepsilon_{12}, \varepsilon_{23}, \varepsilon_{13}, \varepsilon_{123}\}$ of $\boldsymbol{G}(\mathbb{R}^3)$, i.e., one scalar; three orthogonal vectors $\{\varepsilon_i\}$ which can be directly obtained from basis for $\mathbb{R}^3$; three bivectors $\{\varepsilon_{12}, \varepsilon_{23}, \varepsilon_{13}\}$ which are attained from the geometric product between two orthogonal vectors $\{\varepsilon_i\}$ and $\{\varepsilon_j\}$, i.e., $\varepsilon_{ij} = \varepsilon_i \varepsilon_j = \varepsilon_i \wedge \varepsilon_j$ $(i, j = 1, 2, 3,\ i \neq j)$; and one trivector $\{\varepsilon_{123}\}$ is expressed as $\varepsilon_{123} = \varepsilon_1 \varepsilon_2 \varepsilon_3$. In order to introduce geometric product more simply, we consider two specific vectors $s1 = 4\varepsilon_1$ and $s2 = 3\varepsilon_1 + 5\varepsilon_3$. Then,

$$\begin{aligned} s_1 s_2 &= 4\varepsilon_1(3\varepsilon_1 + 5\varepsilon_3) \\ &= 4\varepsilon_1 \cdot (3\varepsilon_1 + 5\varepsilon_3) + 4\varepsilon_1 \wedge (3\varepsilon_1 + 5\varepsilon_3) \\ &= 12 + 20(\varepsilon_1 \wedge \varepsilon_3) \\ &= 12 + 20\varepsilon_{13}. \end{aligned} \tag{7.4}$$

In $\boldsymbol{G}(\mathbb{R}^N)$, a multi-vector A can be expressed as

$$A = \langle A \rangle_0 + \langle A \rangle_1 + \langle A \rangle_2 + \cdots = \sum_k \langle A \rangle_k, \tag{7.5}$$

where $\langle \cdot \rangle_k$ denotes the k-vectors, i.e., $k = 0$ (scalar), $k = 1$ (vectors), $k = 2$ (bivectors), $k = 3$ (trivectors) and so on. Moreover, A can be also expressed as $A = \sum_{k \in D} A_k \varepsilon_k$ with $D = \{0, 1, 2, 3, 12, 13, 23, 123\}$ and $\{A_k\}$ denotes the real-valued coefficient. And the reverse of the a multi-vector A is defined as $\widetilde{A} \triangleq \sum_{k=0}^{N} (-1)^{k(k-1)/2} \langle A \rangle_k$ (for example, when $N = 3$, $\widetilde{A} = \langle A \rangle_0 + \langle A \rangle_1 - \langle A \rangle_2 - \langle A \rangle_3$). The reverse of multi-vector A is somewhat similar to the conjugation of quaternions and complex number.

7.3 Adaptive Filtering Based on Hybrid Geometric Algebra Correntropy Criterion

This section is divided into two main parts; the first part defines the HGAC through the lens of probability density; the second part provides a detailed derivation of the HMGACC algorithm.

7.3.1 Hybrid Geometric Algebra Correntropy Concept

In this part, the HGAC will be proposed in detail, and the HGAC is obtained from the probability view. Here, consider two random multi-vectors, i.e.,

$$M_v = M_0 + \sum_{i=1}^{3} M_i \varepsilon_i + \sum_{j \subset \mathrm{B}} M_j \varepsilon_j + M_{123}\varepsilon_{123}, \tag{7.6}$$

$$N_v = N_0 + \sum_{i=1}^{3} N_i \varepsilon_i + \sum_{j \subset \mathrm{B}} N_j \varepsilon_j + N_{123}\varepsilon_{123}, \tag{7.7}$$

where M_v, N_v are the trivectors (i.e., $M_v, N_v \in \boldsymbol{G}(\mathbb{R}^3)$), $\mathrm{B} = \{12, 13, 23\}$ and M_i, N_i are random variables. When assuming that M_i, N_i have a smooth joint probability density function (PDF), i.e.,$f_{M_0N_0}f_{M_1N_1}\cdots f_{M_{123}N_{123}}(m_0, n_0, \ldots, m_{123}, n_{123})$, the probability density of the event $M_v = N_v$ can be expressed as

$$\begin{aligned} &P_d(M_v = N_v) \\ &= \underbrace{\int_{-\infty}^{\infty} \cdots \int_{-\infty}^{\infty}}_{16} f_{M_0N_0}f_{M_1N_1}\cdots f_{M_{123}N_{123}}(m_0, n_0, \ldots, m_{123}, n_{123}) \\ &\delta(m_0 - n_0)\cdots\delta(m_{123} - n_{123})dm_0dn_0\cdots dm_{123}dn_{123}, \end{aligned} \tag{7.8}$$

where $\delta(\cdot)$ stands for the delta function and 16 denotes the number of integral symbols.

Remake 1 According to [18], the necessary and sufficient condition for $M_v = N_v$ is $M_i = N_i$, $i \in D$.

By the nature of the delta function, we know that the above equation only makes sense when $m_i = n_i$. Therefore, the Eq. (7.8) can be rewritten as

$$\begin{aligned} &P_d(M_v = N_v) \\ &= \underbrace{\int_{-\infty}^{\infty} \cdots \int_{-\infty}^{\infty}}_{8} f_{M_0N_0}f_{M_1N_1}\cdots f_{M_{123}N_{123}}(m_0, n_0, \ldots, m_{123}, n_{123}) \\ &dm_0\cdots dm_{123}. \end{aligned} \tag{7.9}$$

In fact, in specific applications, the true distribution of M_i, N_i, $i \in D$ are not available in advance. Thus, it is necessary here to estimate the joint PDF using a finite number of data samples. By employing the Parzen estimation [17], that is

$$f_{M_1}f_{M_2}\cdots f_{M_T}(m_1, n_2, \ldots, m_T) = \frac{1}{L}\sum_{l=1}^{L}\prod_{t=1}^{T} V_\sigma^M\left(m_t - m_t^{(l)}\right), \tag{7.10}$$

where $m_t^{(l)}$ represents the lth sample for the tth element of the random vector with T-dimensional. $V_\sigma^M(m)$ denotes the mixture Gaussian kernel function that defined as

$$V_\sigma^M(m) = \sum_{i=1}^{2}\alpha_i \exp\left(-\frac{m^2}{2\sigma_i^2}\right), \tag{7.11}$$

where α_i is the proportionality parameter and σ_i is the kernel width. Assuming $T = 16$ and using (7.10), the Eq. (7.9) can be rewritten as

$$P_d(M_v = N_v) = \int_{-\infty}^{\infty}\cdots\int_{-\infty}^{\infty} V_\sigma^M\left(m_0 - m_0^{(l)}\right)V_\sigma^M\left(m_0 - n_0^{(l)}\right)\cdots$$
$$V_\sigma^M\left(m_{123} - m_{123}^{(l)}\right)V_\sigma^M\left(m_{123} - n_{123}^{(l)}\right) dm_0\, dm_1 \cdots dm_{123}. \tag{7.12}$$

Based on *Remake 1*, by employing some algebra, and integral calculations. The above equation can be expressed as

$$P_d(M_v = N_v) = \frac{1}{L}\sum_{l=1}^{L} V_{\sigma\sqrt{2}}^M\left(m_0 - n_0^{(l)}\right)\cdots V_{\sigma\sqrt{2}}^M\left(m_{123}^{(l)} - n_{123}^{(l)}\right). \tag{7.13}$$

The above equation is a general expression of the non-parametric estimation of the probability density for the event $M = N$. Then, the HGAC for M_v, N_v is defined by using the mixture Gaussian kernel as

$$V_\sigma^G(M_v, N_v) = E_{M_vN_v}\left[V_{M_vN_v}(M_v - N_v)\right], \tag{7.14}$$

where

$$V_{M_vN_v}(M_v - N_v) = \sum_{i=1}^{2}\alpha_i \exp\left(-\frac{(M_v - N_v) \times \widetilde{(M_v - N_v)}}{2\sigma_i^2}\right), \tag{7.15}$$

where ($\times$) denotes the scalar product. For two multi-vectors M_v, N_v, the scalar product is defined as

$$M_v \times N_v \triangleq \langle M_v N_v \rangle_0. \tag{7.16}$$

Equation (7.14) can be expressed as Eq. (7.17) via utilizing the joint PDF of M_i, N_i, that is

$$
\begin{aligned}
& V_{\sigma}^{G}(M_v, N_v) \\
& = \int_{-\infty}^{\infty} \cdots \int_{-\infty}^{\infty} f_{M_0 N_0} f_{M_1 N_1} \cdots f_{M_{123} N_{123}}(m_0, n_0, \ldots, m_{123}, n_{123}) \\
& V(m_0 - n_0) \cdots V(m_{123} - n_{123}) dm_0 dm_1 \cdots dm_{123}.
\end{aligned}
\tag{7.17}
$$

By comparing Eqs. (7.8) and (7.17), we can conclude that the HGAC of two multi-vector variables can be expressed using the probability density of the event $M_v = N_v$. Therefore, the non-parameter estimation of HGAC can be calculated as

$$
V_{\sigma}^{G}(M_v, N_v) = \frac{1}{L} \sum_{l=1}^{L} \sum_{i=1}^{2} \alpha_i \exp\left(- \frac{\left(m_0^{(l)} - n_0^{(l)}\right)^2 + \cdots + \left(m_{123}^{(l)} - n_{1213}^{(l)}\right)}{2\sigma_i^2} \right).
\tag{7.18}
$$

Equation (7.18) can be used as a measure of completeness for two random multi-vector-valued variables. In addition, the HGAC is also symmetric due to $V_{M_v N_v}(M_v - N_v) = V_{N_v M_v}(N_v - M_v)$.

7.3.2 System Identification Model

Consider a system identification problem, a linear multi-vector valued model can be described as

$$
d_m = \boldsymbol{x}_m \boldsymbol{\Omega}^{\text{opt}} + \eta_m = \sum_{s=1}^{S} x_{m-s+1} \boldsymbol{\Omega}_s^{\text{opt}} + \eta_m,
\tag{7.19}
$$

where $\boldsymbol{x}_m = [x_m, x_{m-1}, x_{m-S+1}]$ denotes the input signal with length S and the weight vector is $\boldsymbol{\Omega}_m = [\Omega_1, \Omega_2, \ldots, \Omega_S]^{\text{T}}$, where $(\cdot)^{\text{T}}$ is for transpose operation. The unknown weight multi-vector array is $\boldsymbol{\Omega}^{\text{opt}} = \left[\Omega_1^{\text{opt}}, \Omega_2^{\text{opt}}, \ldots, \Omega_S^{\text{opt}}\right]^{\text{T}}$ need to be estimated, where the kth entry Ω_k^{opt} has multi-vector-value. By the definition of multi-vector in the previous section, Ω_k^{opt} can also be expressed as $\Omega_k^{\text{opt}} = \sum_{k \in D} \overline{\Omega}_k^{\text{opt}} \varepsilon_m$, where $\overline{\Omega}_m^{\text{opt}}$ denotes the real-valued coefficient. The mth entries x_{m-S+1} of $\boldsymbol{x}_m$, Ω_m, and Ω_m^{opt} are the multi-vector-valued components. d_m is the desired signal of the system, and η_m represents the measurement noise at time m. The output signal is expressed as

$$
y_m = \sum_{s=1}^{S} x_{m-s+1} \boldsymbol{\Omega}_{s,m}.
\tag{7.20}
$$

According to (7.5), the ith entry x_{m-S+1} of $\boldsymbol{x}_m$ can also be written as

$$x_{m-s+1} = \sum_{k \in D} \overline{x}_{k,m-s+1} \varepsilon_k, \tag{7.21}$$

where $\overline{x}_{k,m-s+1}$ is the real-valued coefficients. And $\boldsymbol{x}_m$ can be expressed with $\overline{\boldsymbol{x}}_m$ via employing Eq. (7.21) as

$$\boldsymbol{x}_m = \sum_{k \in D} \overline{\boldsymbol{x}}_{k,m} \varepsilon_k, \tag{7.22}$$

where $\overline{\boldsymbol{x}}_{k,m} = [\overline{x}_{k,m}, \overline{x}_{k,m-1}, \overline{x}_{k,m-S+1}]$ is the real-valued coefficients. Based on (7.20–7.22), the weight vector $\boldsymbol{\Omega}_m$ and the output signal y_m can be rewritten as

$$\boldsymbol{\Omega}_m = \sum_{k \in D} \overline{\Omega}_{k,m} \varepsilon_k, \tag{7.23}$$

$$y_m = \sum_{k \in D} \overline{y}_{k,m} \varepsilon_k, \tag{7.24}$$

where $\overline{\Omega}_{k,m}$ represents the real-valued coefficients, $\overline{y}_{k,m}$ is calculated as

$$\overline{y}_{k,m} = \left(\sum_{s=1}^{S} x_{m-s+1} \boldsymbol{\Omega}_m^s \right)_k = \sum_{p,q \in D} \left(\overline{\boldsymbol{x}}_{p,m} \overline{\Omega}_{q,m} \right) \left\langle \varepsilon_k \widetilde{\varepsilon}_q \widetilde{\varepsilon}_p \right\rangle. \tag{7.25}$$

Then, y_m can be re-expressed as

$$y_m = \sum_{k,p,q \in D} \left(\overline{\boldsymbol{x}}_{p,m} \overline{\Omega}_{q,m} \right) \left\langle \varepsilon_k \widetilde{\varepsilon}_q \widetilde{\varepsilon}_p \right\rangle \varepsilon_k. \tag{7.26}$$

7.3.3 Deviation of the HMGACC Adaptive Algorithm

First, the multi-vector-valued error e_m between y_m and d_m is calculated as

$$e_m = \sum_{k \in D} \overline{e}_{k,m} \varepsilon_k = d_m - y_m. \tag{7.27}$$

Recalling Eq. (7.18), the HGAC-based cost function is given by the HAC between the multi-vector-valued output signal y_m and the multi-vector-valued desired signal d_m. The optimal solution for the weight vector can be obtained by minimizing the HGAC-based cost function (i.e., Eq. (7.28)).

In other word, the optimal solution for $\mathbf{\Omega}$ can be calculated by maximizing the cost function J_{HGAC}. This optimization principle is denoted by the hybrid maximum GAC criteria (HMGACC).

$$\begin{aligned} J_{\text{HGAC}} &= 1 - \frac{1}{L}\sum_{m=1}^{L}\sum_{i=1}^{2}\alpha_i \exp\left(-\frac{(d_m - y_m)\times\widetilde{(d_m - y_m)}}{2\sigma_i^2}\right) \\ &= 1 - \frac{1}{L}\sum_{m=1}^{L}\sum_{i=1}^{2}\alpha_i \exp\left(-\frac{|e_m|^2}{2\sigma_i^2}\right). \end{aligned} \tag{7.28}$$

Under the HMGACC, by approximating the sum using current values, we can obtain the instantaneous cost function, i.e.,

$$J_{\text{MGAC}} = 1 - \sum_{i=1}^{2}\alpha_i exp\left(-\frac{|e_m|^2}{2\sigma_i^2}\right). \tag{7.29}$$

Using the stochastic gradient method, the adaptive algorithm based on MMGACC is given by

$$\mathbf{\Omega}_{m+1} = \mathbf{\Omega}_m + \mu\frac{\partial J_{\text{MGAC}}}{\partial\mathbf{\Omega}} = \mathbf{\Omega}_m - \mu\sum_{i=1}^{2}\frac{\alpha_i}{\sigma_i^2}\text{expexp}\left(-\frac{|e_m|^2}{2\sigma_i^2}\right)\frac{|e_m|^2}{\partial\mathbf{\Omega}}, \tag{7.30}$$

where μ is the step size, and $|e_m|^2$ can be rewritten as

$$|e_m|^2 = e_m \times \widetilde{e_m} == \sum_{k\in D}\left(\overline{d}_{k,m} - \overline{y}_{k,m}\right)^2, \tag{7.31}$$

thus, there are

$$\frac{|e_m|^2}{\partial\mathbf{\Omega}} = \frac{\sum_{k\in D} e_{k,m}^2}{\partial\mathbf{\Omega}} = -2\sum_{k\in D}\overline{e}_{k,m}\left(\frac{\overline{y}_{k,m}}{\partial\mathbf{\Omega}}\right). \tag{7.32}$$

According to [19], we have

$$\frac{\overline{y}_{k,m}}{\partial\mathbf{\Omega}} = \sum_{r\in D}\frac{\overline{y}_{k,m}}{\partial\mathbf{\Omega}_r}\varepsilon_r, \tag{7.33}$$

and recalling Eq. (7.25), then substituting (7.25) into (7.33) yields

$$\frac{\overline{y}_{k,m}}{\partial\mathbf{\Omega}} = \frac{\sum_{p,q,r\in D}\left(\overline{\boldsymbol{x}}_{p,m}\overline{\Omega}_{q,m}\right)\left\langle\varepsilon_k\widetilde{\varepsilon}_q\widetilde{\varepsilon}_p\right\rangle\varepsilon_r}{\partial\mathbf{\Omega}_r} = \sum_{p,q,r\in D}\delta_{r,q}\overline{\boldsymbol{x}}_{p,m}\left\langle\varepsilon_k\widetilde{\varepsilon}_q\widetilde{\varepsilon}_p\right\rangle\varepsilon_r, \tag{7.34}$$

by employing $\widetilde{A} = \langle A\rangle_0 + \langle A\rangle_1 - \langle A\rangle_2 - \langle A\rangle_3$ and Eq. (7.22), the reverse of $\boldsymbol{x}_m$ can be calculated as

$$\widetilde{\boldsymbol{x}_m} = \sum_{p\in D} \overline{\boldsymbol{x}}_{p,m}\widetilde{\varepsilon_p}, \tag{7.35}$$

Substituting (7.35) into (7.34) yields

$$\frac{\overline{y}_{k,m}}{\partial \boldsymbol{\Omega}} = \sum_{q\in D} \left\langle \varepsilon_k \widetilde{\varepsilon_q} \widetilde{\boldsymbol{x}_m} \right\rangle \varepsilon_q, \tag{7.36}$$

using the calculation rules in the literature [20], there are

$$\left\langle \varepsilon_k \widetilde{\varepsilon_q} \widetilde{\boldsymbol{x}_m} \right\rangle = \langle \varepsilon_k \widetilde{\boldsymbol{x}_m} \rangle_q, \tag{7.37}$$

substituting the above equation into (36) yields

$$\frac{\overline{y}_{k,m}}{\partial \boldsymbol{\Omega}} = \sum_{q\in D} \langle \varepsilon_k \widetilde{\boldsymbol{x}_m} \rangle_q \varepsilon_q = \varepsilon_k \widetilde{\boldsymbol{x}_m}. \tag{7.38}$$

Substituting the above equation into (7.32), the recursion of the proposed HMGACC algorithm can be expressed as

$$\boldsymbol{\Omega}_{m+1} = \boldsymbol{\Omega}_m + \sum_{i=1}^{2} \mu_i \exp\left(-\frac{|e_m|^2}{2\sigma_i^2}\right) \sum_{k\in D} \overline{e}_{k,m}\varepsilon_k \widetilde{\boldsymbol{x}_m}, \tag{7.39}$$

where $\mu_i = \mu\alpha_i / \sigma_i^2$ is the step size, which is controlled by the proportionality parameter α_i, the kernel-width σ_i, and the original step-size μ.

7.4 Computer Simulation Examples

In this section, the presented HMGACC algorithm is compared with other existing GA based algorithms in different noise environments. All the entries of the unknown weight vector $\boldsymbol{\Omega}^{\mathrm{opt}}$, i.e., $\boldsymbol{\Omega}_k^{\mathrm{opt}}$, are multi-vectors, and $\boldsymbol{\Omega}_k^{opt}$ is given by $\boldsymbol{\Omega}_k^{\mathrm{opt}} = [0.5 - 3.1\varepsilon_1 + 1.7\varepsilon_2 + 2\varepsilon_2 + 0.71\varepsilon_{12} + 1.3\varepsilon_{23} - 4.5\varepsilon_{31} + 3I]$. The blade coefficients of each element of the input signal array $\boldsymbol{x}_m$ are drawn from a zero-mean white Gaussian noise with variance $\delta_x^2 = 1$. And to better test the proposed algorithm performance, the noise is generated in two ways: one is generated by the noise model I, i.e., $\eta_m = \eta_{\theta 1,m} + a_m\eta_{Im1,m}$. Another is generated by the model II, i.e., $\eta_m = (1 - b_m)\eta_{\theta 2,m} + b_m\eta_{Im2,m}$, where $\eta_{\theta 1,m}$, $\eta_{02,m}$ denote the background noise, the blade coefficients of each element of $\eta_{\theta 1,m}$, $\eta_{\theta 2,m}$ are generated by zero-mean

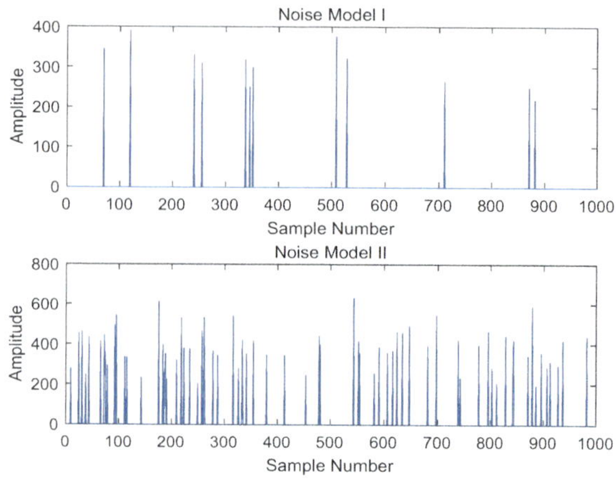

Fig. 7.1 Comparison of noise generated by two models

white Gaussian noise with variances 0.01 and 0.1, respectively. a_m is a Bernoulli-Gaussian process, b_m is a binary independent and identically distributed process with $\Pr\{a_m = 1\} = p_1$, $\Pr\{a_m = 0\} = 1 - p_1$, $\Pr\{b_m = 1\} = p_2$, $\Pr\{b_m = 0\} = 1 - p_2$, where $0 \le p_1, p_2 \le 1$ are the probability of occurrence of impulse noise; $\eta_{Im1,m}$ and $\eta_{Im2,m}$ denote the impulse noise, the blade coefficients of each element of $\eta_{Im1,m}$ and $\eta_{Im2,m}$ are generated by zero-mean white Gaussian noise with variances 10,000 and 20,000, respectively. The noise processes $\eta_{\theta 1,m}$ and $\eta_{Im1,m}$ are mutually independent and they are both independent of a_m; $\eta_{\theta 2,m}$ and $\eta_{Im2,m}$ are mutually independent and they are both independent of b_i. The mean square deviation (MSD) is defined as $\text{MSD}_m = 10\log_{10}(||\boldsymbol{\Omega}^{\text{opt}} - \boldsymbol{\Omega}_m||^2)$, and used as the performance measures of the algorithms. All learning curves are obtained by averaging over 50 independent runs.

7.4.1 Comparisons of Different Algorithms

We choose $p_1 = 0.01$,$p_2 = 0.07$ to generate the noises and the noise model is shown in Fig. 7.1 for testing the performance of the proposed algorithm with other algorithms and the results are shown in Figs. 7.2 and 7.3. As can be seen that although the MGACC algorithms also maintain a good performance under the interference of the two different noises, the proposed algorithm is able to perform better than the other algorithms, i.e., faster convergence and lower steady-state error.

7.4.2 Tracking Performance Comparison

In this section, the unknown weights change abruptly at 1000 iterations, i.e., $\boldsymbol{\Omega}_k^{\text{opt}} = -\boldsymbol{\Omega}_k^{\text{opt}}$. The result of the tracking learning curve is shown in Fig. 7.4, from which

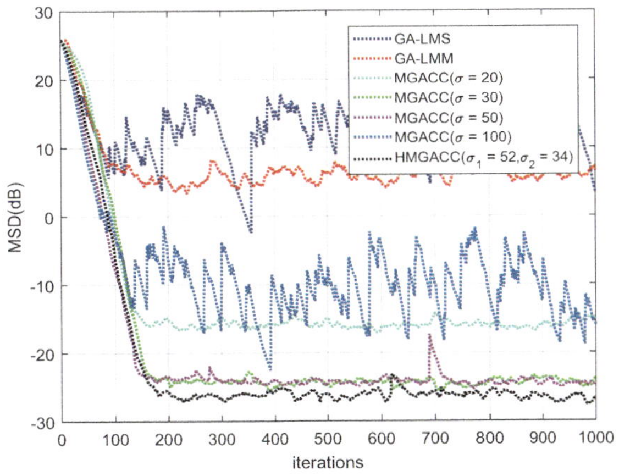

Fig. 7.2 Learning curves between the HMGACC ($\alpha_1 = 0.63, \alpha_2 = 1 - \alpha_1$) and other algorithms with the noise interference generated by the noise model I

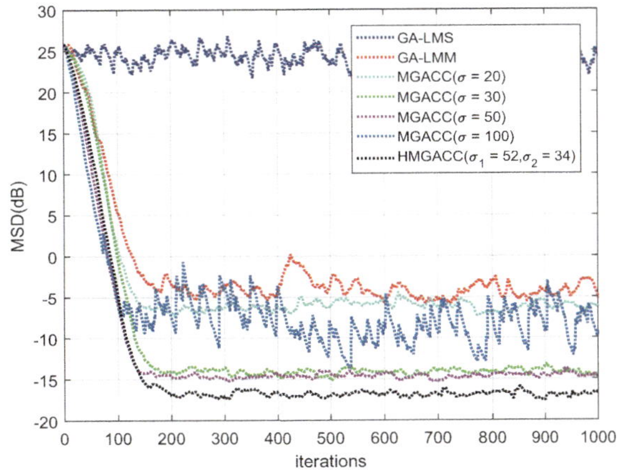

Fig. 7.3 Learning curves between the HMGACC ($\alpha_1 = 0.63, \alpha_2 = 1 - \alpha_1$) and other algorithms with the noise interference generated by the noise model II

it can be seen that the proposed algorithm still maintains good convergence and steady-state performance compared to other algorithms.

7.5 Summary

In this chapter, a hybrid GAC is defined through the lens of probability density, and using this GAC, we propose a hybrid GAC robust adaptive filtering algorithm. This hybrid GAC is more general than the traditional single kernel GAC; and we add a proportionality parameter to control the weights of the different kernel-width correntropy, which increases the flexibility of the correntropy. The step size is also changed to a non-fixed form by the introduction of the proportionality parameter, i.e., the step size used in this chapter is a hybrid step size, which optimizes the

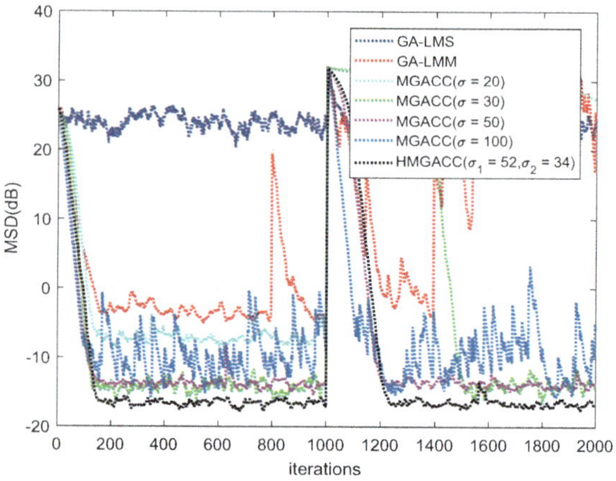

Fig. 7.4 Tracking curves between the HMGACC ($\alpha_1 = 0.63, \alpha_2 = 1 - \alpha_1$) and other algorithms with the noise interference generated by the noise model II

performance of the algorithm better than a fixed step size. The better convergence and steady-state performance of the proposed algorithm in the experimental simulation stage also demonstrates the feasibility of our proposed method. Finally, the concrete application of the HMGACC algorithm will be our future work.

References

1. M. Godavarti, A.O. Hero, Partial update LMS algorithms. IEEE Trans. Signal Process. **53**(7), 2382–2399 (2005)
2. J. Benesty, H. Rey, L.R. Vega, S. Tressens, A nonparametric VSS NLMS algorithm. IEEE Signal Process. Lett. **13**(10), 581–584 (2006)
3. X. Long, H. Zhao, X. Hou, A novel combinatoric correntropy algorithm: properties and its performance analysis. IEEE Trans. Circ. Syst. II Exp. Briefs **69**(12), 5184–5188 (2022)
4. B. Chen, X. Wang, Y. Li, J.C. Principe, Maximum correntropy criterion with variable center. IEEE Signal Process. Lett. **26**(8), 1212–1216 (2019)
5. R. He, B.-G. Hu, W.-S. Zheng, X.-W. Kong, Robust principal component analysis based on maximum correntropy criterion. IEEE Trans. Image Process. **20**(6), 1485–1494 (2011)
6. M. Li, Z. Jing, H. Leung, Robust minimum error entropy based cubature information filter with non-Gaussian measurement noise. IEEE Signal Process. Lett. **28**, 349–353 (2021)
7. H. Zhao, Y. Gao, Y. Zhu, Robust subband adaptive filter algorithms-based mixture correntropy and application to acoustic echo cancellation. IEEE Trans. Audio Speech Lang. Process. **31**, 1223–1233 (2023)
8. H. Zhao, B. Tian, Robust power system forecasting-aided state estimation with generalized maximum mixture correntropy unscented Kalman filter. IEEE Trans. Instrum. Meas. **71**, 1–10 (2022)
9. X. Hou, H. Zhao, X. Long, Graph diffusion kernel maximum correntropy criterion over sensor network and its performance analysis. IEEE Sensors J. **23**(13), 14583–14591 (2023)
10. H. Zhao, Z. Cao, Robust generalized maximum Blake–Zisserman total correntropy adaptive filter for generalized Gaussian noise and noisy input. IEEE Trans. Syst. Man Cybern. https://doi.org/10.1109/TSMC.2023.3287406
11. W.B. Lopes, A. Al-Nuaimi, C.G. Lopes, Geometric-algebra LMS adaptive filter and its application to rotation estimation. IEEE Signal Process. Lett. **23**(6), 858–862 (2016)

12. W. Wang, K. Doğançay, Transient performance analysis of geometric algebra least mean square adaptive filter. IEEE Trans. Circ. Syst. II Exp. Briefs **68**(8), 3027–3031 (2021)
13. W.B. Lopes, C.G. Lopes, Geometric-algebra adaptive filters. IEEE Trans. on Signal Process. **67**(14), 3649–3662 (2019)
14. B. Chen, X. Wang, N. Lu, S. Wang, J. Cao, J. Qin, Mixture correntropy for robust learning. Pattern Recog. **79**, 318–327 (2018)
15. S. Lv, H. Zhao, X, He, Geometric algebra based least mean m-estimate robust adaptive filtering algorithm and its transient performance analysis, Signal Process. **189** (2021)
16. W. Wang, H. Zhao, X. Zeng, Geometric algebra correntropy: definition and application to robust adaptive filtering. IEEE Trans. Circ. Syst. II Exp. Briefs **67**(6), 1164–1168 (2020)
17. W. Liu, P.P. Pokharel, J.C. Principe, Correntropy: properties and applications in non-Gaussian signal processing. IEEE Trans. Signal Process. **55**(11), 5286–5298 (2007)
18. E.J. Bayro-Corrochano, N. Arana-Daniel, Clifford support vector machines for classification, regression, and recurrence. IEEE Trans. Neural Netw. **21**(11), 1731–1746 (2010)
19. C. Perwass, H. Edelsbrunner, L. Kobbelt, K. Polthier, Geometric algebra with applications in engineering, vol. 4 (Springer, 2009)
20. E. Hitzer, Algebraic foundations of split hypercomplex nonlinear adaptive filtering. Math. Methods Appl. Sci. **36**(9), 1042–1055 (2013)

Chapter 8
Kernel Generalized Maximum Correntropy with Variable Centers

Recently, the kernel adaptive filtering algorithm based on the generalized maximum correntropy criterion (GMCC) has attracted a lot of attention due to its excellent nonlinear modeling ability and robustness to non-Gaussian noise. However, since the default center of the GMCC is located at zero, it leads to a significant deterioration of the algorithm performance in the nonzero-mean noise environment. In order to overcome the above-mentioned problem, this paper proposes a kernel adaptive filtering algorithm based on the GMCC with variable center (GMCC-VC) called kernel generalized maximum correntropy criterion with variable center (KGMCC-VC). The center of the proposed algorithm can be located at any position, which reduces the impact of nonzero-mean noise on the performance of the algorithm to some extent. In addition, the online vector quantization (VQ) method is introduced to the KGMCC-VC algorithm for suppressing the increasing network size. The probability of divergence, the mean-square steady-state behavior, and the convergence condition of the KGMCC-VC algorithm is also obtained under some assumptions. The potential relationship between the GMCC-VC and the generalized minimum error entropy (GMEE) criteria is discussed for the first time in the paper. Finally, the simulation results verify the superiority of the proposed algorithm in modeling nonlinear systems and the correctness of the theoretical model.

8.1 Introduction

Kernel methods have been developed comprehensively as a powerful nonlinear modeling tool in the fields of pattern recognition and machine learning. It has been successfully applied in the fields of classification and feature extraction, such as support vector machines (SVM) [1, 2], kernel principal component analysis (KPCA) [3], etc. The main idea of kernel methods is to transform the input signal by feature mapping into the high-dimensional feature space induced by the Mercer kernel so

H. Zhao et al., *Information Theoretic Learning-Based Filter*,
https://doi.org/10.1007/978-3-032-29623-8_8

that the inner product operation in the feature space can be replaced by a kernel function [4]. The transformed data is then processed with a suitable linear algorithm to construct a nonlinear relationship between the input and output efficiently. Using kernel methods, we can extend classical linear algorithms, including adaptive filtering, to the reproducing kernel Hilbert space (RKHS) to solve nonlinear signal processing problems.

In recent years, the kernel adaptive filtering algorithms have attracted a lot of attention due to their ability to adapt to changes in the environment in time, which largely enhances the adaptability of the models compared to other offline kernel learning algorithms such as kernel extreme learning machine (KELM) [5] and kernel ridge regression (KRR) [6]. The kernel adaptive filtering algorithm is a family of nonlinear filtering algorithms whose computational complexity is only related to the number of input samples. Therefore, it effectively overcomes the dilemma that the computational complexity of the traditional Volterra filter [7] grows exponentially with the increase of the filtering order. In addition, the kernel adaptive filtering algorithm inherits the simplicity of linear filter structure, so it is widely used in the fields of nonlinear system modeling, time series prediction [8, 9], etc. The typical kernel adaptive filtering algorithms include kernel least mean square (KLMS) [10–12], kernel affine projection algorithm (KAPA) [13], kernel recursive least square (KRLS) [14, 15].

The above-mentioned kernel adaptive filtering algorithms are constructed based on the mean square error (MSE) criterion. The MSE criterion is the optimal criterion for the linear adaptive filtering algorithm under the Gaussian noise assumption. However, the performance of the adaptive filtering algorithm based on the MSE criterion deteriorates significantly when the signal contains the non-Gaussian noise as it contains only the second-order statistical information of the error. For this problem, many criteria based on information learning theory have been proposed and the corresponding kernel adaptive filtering algorithms have been developed, such as the kernel maximum correntropy criterion (KMCC) [16] and kernel minimum error entropy (KMEE) algorithms [17]. Adaptive filtering algorithms based on MCC and MEE criteria usually perform more robustly than MSE-based adaptive filtering algorithms because higher-order error information is involved [18]. However, the Gaussian kernel function is not always an optimal choice. Therefore, combined with the generalized Gaussian density (GGD) function, some generalized maximum correntropy criterion (GMCC) [19, 20] and generalized minimum error entropy (GMEE) [21] criteria are proposed to improve the learning performance. Unfortunately, neither the MCC-based nor the GMCC-based kernel adaptive filtering algorithm can achieve the desired results when the noise mean is not zero. Because the centers of both the Gaussian kernel and the generalized Gaussian kernel functions are always located at zero, they are not aligned with the nonzero-mean noise. Later, the MCC with variable center (MCC-VC) was proposed to deal with the case of nonzero-mean noise, and the corresponding kernel adaptive filtering algorithm was derived [22]. To extend the type of kernel functions of correntropy, the GMCC with variable center (GMCC-VC) is proposed [23]. The center location of the GMCC-VC criterion can be used to capture the distribution information of the sampling error in the GMEE criterion. As far as the GMEE-based kernel adaptive filtering algorithm is concerned, its computational complexity is comparable to that of the KAPA algorithm. Compared with the GMEE criterion, although GMCC-VC loses the error scatter information, it can significantly decrease the computational complexity of the adaptive filtering algorithm based on the GMEE criterion. Motivated by the excellent properties of the GMCC-VC, a kernel generalized maximum correntropy criterion with variable center (KGMCC-VC) algorithm is proposed in this paper. Considering the problem that the computational burden of the kernel adaptive filtering algorithm increases with the number of data, this paper further introduces the sparsification method of online vector quantization (VQ) [24] to suppress the growth of computational cost, resulting in the quantized KGMCC-VC (QKGMCC-VC). The simulation results verify the superiority of the KGMCC-VC in nonzero-mean noise environment.

In addition, the performance analysis is an essential issue for kernel adaptive filtering algorithms. There are three prevailing analytical approaches: predetermined dictionaries [25, 26], random Fourier feature [27] and energy conservation relation [28]. However, in the traditional sense, the predetermined dictionary analysis is not essentially a kernel adaptive filtering algorithm analysis. It is closer to a linear filter after feature extraction of the input signal and cannot be used directly to predict the behavior of the standard kernel adaptive filtering algorithm. Analysis methods using random Fourier feature often require a large number of features, and their results are not precise, as shown in [27]. Therefore, the mean-square steady-state behavior of the KGMCC-VC algorithm and the convergence conditions are analyzed in this paper by means of energy conservation relations. The correctness of the theoretical analysis model is verified in the simulation section.

8.2 Generalized Maximum Correntropy with Variable Centers

The correntropy is a nonlinear measure of the similarity between two random variables A and B [29], which is defined as

$$V(A, B) = E[k_\delta(A, B)] = \int k_\delta(a, b)\mathrm{d}F_{A,B}(a, b) \tag{8.1}$$

where $k_\delta(\cdot, \cdot)$ is a shift-invariant Mercer kernel controlled by the kernel bandwidth δ, $E(\cdot)$ is the expectation operator, and $F_{A,B}(a, b)$ is the joint distribution function of A and B.

The maximum correntropy criterion usually adopts the Gaussian kernel, while the generalized maximum correntropy criterion (GMCC) regards the generalized Gaussian density (GGD) function as the kernel function [19]. The generalized maximum correntropy with zero-mean is expressed as

$$V_{\zeta,\theta}(A, B) = \int G_{\zeta,\theta}(a, b)\mathrm{d}F_{A,B}(a, b) \tag{8.2}$$

where

$$G_{\zeta,\theta}(a, b) = \frac{\zeta}{2\theta\Gamma(1/\zeta)} \exp\left(-\left|\frac{a-b}{\theta}\right|^{\zeta}\right) = \gamma_{\zeta,\theta}\exp\left(-\lambda|e|^{\zeta}\right) \tag{8.3}$$

where $e = a - b$, $\Gamma(\cdot)$ is the gamma function, $\theta > 0$ is the scale parameter, $\zeta > 0$ is the shape parameter, $\lambda = 1/\theta^{\zeta}$ is the kernel parameter, and $\gamma_{\zeta,\theta}$ is the normalization constant.

The KGMCC is a kernel adaptive filtering algorithm using the GMCC as the cost function. Since the definition of the joint distribution function is elusive in (8.2), the cost function of the KGMCC algorithm at discrete time m is simply approximated as

$$J_{\mathrm{GKMCC}} = \gamma_{\zeta,\theta}\exp\left(-\lambda|e(m)|^{\zeta}\right) \tag{8.4}$$

where $e(m)$ is the prediction error. The KGMCC algorithm using gradient descent method is summarized in Algorithm 8.1 and the detailed derivation can be found in the literature [30].

Algorithm 8.1: Pseudo code for KGMCC algorithm

Input: Kernel bandwidth σ ; Step-size η ; Kernel parameter λ ; Shape parameter ζ
Output: Coefficient vector $\boldsymbol{\alpha}$
/* Initialization */
$m \leftarrow 1$, $c(1) \leftarrow y(1)$, $\alpha(1) \leftarrow 0$
/* Computation */
while $m \le N$ **do**
  $m \leftarrow m+1$
  $\hat{d}(m) \leftarrow \sum_{i=1}^{m-1} \alpha_i(m-1)k_{\sigma}\left(x(i), x(m)\right)$
  $e(m) \leftarrow d(m) - \hat{d}(m)$
  $g(e(m)) \leftarrow \exp\left(-\lambda|e(m)|^{\zeta}\right)|e(m)|^{\zeta-1}\operatorname{sign}\left(e(m)\right)$
  $\alpha(m) \leftarrow \left\{\alpha(m-1), \eta g\left(e(m)\right)\right\}$
  end
  $m \leftarrow m+1$
end
return $\boldsymbol{\alpha}$

Since the center of the GGD function is always located at zero, this may result in performance degradation of the GMCC-based algorithm in nonzero-mean noise environment. The GMCC with variable center, similar to the MCC-VC [22], is defined as

$$G_{\zeta,\theta}(e-c)=\gamma_{\zeta,\theta}\exp(-\lambda|e-c|^{\zeta}) \tag{8.5}$$

where $c \in \mathbb{R}$ denotes the center location.

Remark 8.1 Note that the above definition will degrade to the MCC-VC when $\zeta = 2$, the GMCC when $c = 0$, the MCC when $c = 0$ and $\zeta = 2$. In addition, the generalized maximum correntropy with center c consists of the higher-order absolute moments of the error e about the center c.

In practice, only a limited number of samples of random variables A and B are available. Therefore, the sample mean estimator of the GMCC-VC is

$$\hat{V}_{\zeta,\theta}(e)=\frac{1}{N}\sum_{i=1}^{N}G_{\zeta,\theta}(e(i)-c) \tag{8.6}$$

where N is the sample number. On the other hand, the GMEE as shown in [21] can be defined as

$$\hat{V}_{\zeta,\theta}(e)=\frac{1}{N^2}\sum_{i=1}^{N}\sum_{j=1}^{N}G_{\zeta,\theta}(e(i)-e(j)) \tag{8.7}$$

By combining (8.6) and (8.7), it can be seen that the GMCC-VC criterion can be regarded as making the error close to the central position c, while GMEE makes the error distribution more concentrated. This variable center strategy establishes an intrinsic connection between the GMCC-VC and GMEE criteria.

8.3 Kernel Generalized Maximum Correntropy with Variable Centers

Now consider a supervised machine learning problem in which we need to establish a mapping function $f:\mathbb{R}^L \rightarrow \mathbb{R}$ within the dataset $\mathcal{S}=\{\boldsymbol{x}(m), d(m)\}_{m=1}^{N}$, where $\boldsymbol{x}(m) \in \mathbb{R}^L$ is input vector at discrete time m, $d(m) = f(\boldsymbol{x}(m)) + v(m) \in \mathbb{R}$ is desired output, and $v(m) \in \mathbb{R}$ is additive noise. When the mapping function f is nonlinear, which means that the underlying system is nonlinear, the traditional linear adaptive filtering algorithm may not achieve the expected results. Due to the strong nonlinear mapping capacity of kernel methods, the nonlinear mapping function f can be constructed by linear adaptive filtering algorithms in a high-dimensional feature space $\mathbb{F}$. Moreover, the mapping function f is usually represented by the weight vector $\boldsymbol{\omega}$ in kernel adaptive algorithms. The mapping function f (weight vector $\boldsymbol{\omega}$) according to the represented theorem is expressed as

$$f=\sum_{m=1}^{n}\alpha(m)\boldsymbol{\psi}(\boldsymbol{x}(m))=\sum_{m=1}^{n}\alpha(m)k_{\sigma}(\boldsymbol{x}(m),\cdot)=\boldsymbol{\omega} \tag{8.8}$$

where $\{\alpha(m)\}$ are expansion coefficients, $\boldsymbol{\psi} : \mathbb{R}^L \rightarrow \mathbb{F}$ is the feature mapping, $k_\sigma(\cdot, \cdot)$ is kernel function of the kernel methods, and σ is kernel bandwidth. The commonly used Gaussian kernel is

$$k_\sigma(\boldsymbol{x}(i), \boldsymbol{x}(j)) = \exp\left(-\frac{\|\boldsymbol{x}(i) - \boldsymbol{x}(j)\|^2}{2\sigma^2}\right) \tag{8.9}$$

Remark 8.2 It is worth noting that the kernel function of correntropy and the kernel function of the kernel methods have different meanings: the former emphasizes the similarity measure in RKHS, while the latter focuses on simplifying the complicated operations in the high-dimensional feature space. In other words, the latter can replace the inner product of two feature vectors $\boldsymbol{\psi}(\boldsymbol{x}(i))$ and $\boldsymbol{\psi}(\boldsymbol{x}(j))$ by a simple operation, i.e.,

$$k_\sigma(\boldsymbol{x}(i), \boldsymbol{x}(j)) = \boldsymbol{\psi}(\boldsymbol{x}(i))^{\mathrm{T}} \boldsymbol{\psi}(\boldsymbol{x}(j))$$

For the KGMCC-VC algorithm, the optimal weight vector of the filter can be obtained by minimizing the following cost function

$$J_{\text{GKMCC-VC}} = \gamma_{\zeta,\theta} - E\left[G_{\zeta,\theta}(e(m) - c)\right] = \gamma_{\zeta,\theta}\left\{1 - E\left[\exp\left(-\lambda|e(m) - c|^\zeta\right)\right]\right\} \tag{8.10}$$

where $e(m)$ is the prediction error

$$e(m) = d(m) - \boldsymbol{\omega}(m-1)^{\mathrm{T}} \boldsymbol{\psi}(\boldsymbol{x}(m)) \tag{8.11}$$

Combined with the stochastic gradient method, the weight update equation of the KGMCC-VC algorithm is represented as

$$\boldsymbol{\omega}(m+1) = \boldsymbol{\omega}(m) + \mu \frac{\partial}{\partial \boldsymbol{\omega}(m)} \exp\left(-\lambda|e(m) - c|^\zeta\right) = \boldsymbol{\omega}(m) + \eta g(e(m)) \boldsymbol{\psi}(\boldsymbol{x}(m)) \tag{8.12}$$

where $g(e(m)) = \exp\left(-\lambda|e(m) - c|^\zeta\right)|e(m) - c|^{\zeta-1}\text{sign}(e(m) - c)$, $\eta = \mu\lambda\zeta$ is the step-size parameter.

Remark 8.3 As one can observe, the above weight update equation will degrade to different kernel adaptive filtering algorithms: (1) the KMC-VC algorithm when $\zeta = 2$; (2) the KGMCC algorithm when $c = 0$; (3) the KMCC algorithm when $c = 0$ and $\zeta = 2$; (4) the KLMP algorithm when $\lambda \rightarrow 0_+$ and $c = 0$; (5) the KLMS algorithm when $\lambda \rightarrow 0_+$, $c = 0$ and $\zeta = 2$. Moreover, the GMCC-VC is similar to the GMEE criterion, in other words, c can be considered as the mean (median) value of the last N sampling errors at time m. Compared with the GMEE-based kernel adaptive filtering algorithm, the KGMCC-VC algorithm does not need to repeatedly calculate the error information of the last N data, thus greatly reducing the computational and storage costs. Although the scatter information of the error distribution is lost, the computational complexity of KGMCC-VC is similar to the KLMS algorithm significantly lower than that of GKMEE.

When a new input data $\boldsymbol{x}(m)$ arrives, the output of the kernel adaptive filter can be represented as an inner product of the new input and all previous weight vectors.

$$d(m+1) = \boldsymbol{\omega}(m+1)^{\mathrm{T}} \boldsymbol{\psi}(x(m+1)) = \eta \sum_{i=1}^{m} g(e(i)) k_{\sigma}(\boldsymbol{x}(i), \boldsymbol{x}(m+1)) \quad (8.13)$$

Algorithm 8.2: Pseudo code for KGMCC-VC algorithm

Input: Kernel bandwidth σ ; Step-size η ; Kernel parameter λ ; Shape parameter ζ ; Sample number N;
Output: Coefficient vector $\boldsymbol{\alpha}$

/* Initialization */
$m \leftarrow 1$, $c(1) \leftarrow y(1)$, $\alpha(1) \leftarrow 0$
/* Computation */
while $m \le N$ **do**
  $m \leftarrow m+1$
  $\hat{d}(m) \leftarrow \sum_{i=1}^{m-1} \boldsymbol{\alpha}_i (m-1) k_{\sigma}(x(i), x(m))$
  $e(m) \leftarrow d(m) - \hat{d}(m)$
  $\epsilon(m) \leftarrow e(m) - c(m-1)$
  $g(e(m)) \leftarrow \exp\left(-\lambda |\epsilon(m)|^{\zeta}\right) |\epsilon(m)|^{\zeta-1} \operatorname{sign}(\epsilon(m))$
  $\boldsymbol{\alpha}(m) \leftarrow \{\boldsymbol{\alpha}(m-1), \eta g(e(m))\}$
  /* Adapt c */
  if $m < N$ **then**
    $c(m+1) \leftarrow 0$
  else
    $c(m+1) \leftarrow \text{median}\{e(m-N+1), \cdots, e(m)\}$
  end
end
return $\boldsymbol{\alpha}$

The value of the free parameter c in the KGMCC-VC algorithm mainly captures the mean property of the noise, which significantly affects the convergence and robustness performance in the presence of nonzero-mean noise. In order to obtain the appropriate values of the free parameter, the following objective functions need to be solved.

Algorithm 8.3: Pseudo code for QKGMCC-VC algorithm

Input: Kernel bandwidth σ ; Step-size η ; Kernel parameter λ ; Shape parameter ζ ; Sample number N; Quantization size ρ
Output: Coefficient vector $\boldsymbol{\alpha}$
/* Initialization */
$m \leftarrow 1$, $c(1) \leftarrow y(1)$, $M \leftarrow 1$, $\mathcal{X}(1) \leftarrow \mathbf{x}(1)$, $\alpha(1) \leftarrow 0$
/* Computation */
while $m \leq N$ **do**
  $m \leftarrow m+1$
  $\hat{d}(m) \leftarrow \sum_{i=1}^{M} \boldsymbol{\alpha}_i(m-1) k_\sigma\left(\mathcal{X}_i(m-1), x(m)\right)$
  $e(m) \leftarrow d(m) - \hat{d}(m)$
  $\epsilon(m) \leftarrow e(m) - c(m-1)$
  $g(e(m)) \leftarrow \exp\left(-\lambda |\epsilon(m)|^{\zeta}\right) |\epsilon(m)|^{\zeta-1} \operatorname{sign}\left(\epsilon(m)\right)$
  if $\operatorname{dis}\left(x(m), \mathcal{X}(m-1)\right) < \rho$ **then**
    $\mathcal{X}(m) \leftarrow \mathcal{X}(m-1)$
    $i \leftarrow \arg\min_{1 \leqslant i \leqslant M} \| x(m) - \mathcal{X}_i(m-1) \|$
    $\boldsymbol{\alpha}_i(m) \leftarrow \boldsymbol{\alpha}_i(m-1) + \eta g(e(m))$
  else
    $\mathcal{X}(m) \leftarrow \{\mathcal{X}(m-1), \mathbf{x}(m)\}$
    $\boldsymbol{\alpha}(m) \leftarrow \{\boldsymbol{\alpha}(m-1), \eta g\left(e(m)\right)\}$
    $M \leftarrow M+1$
  end
  /* Adapt c */
  if $m < N$ **then**
    $c(m+1) \leftarrow 0$
  else
    $c(m+1) \leftarrow \operatorname{median}\{e(m-N+1), \cdots, e(m)\}$
  end
end
return $\boldsymbol{\alpha}$

$$c(m) = \arg\min_{c \in C} \left\{ \gamma_{\zeta,\theta} - E\left[G_{\zeta,\theta}(e(m) - c) \right] \right\} \tag{8.14}$$

where C denotes the admissible sets of c. To simplify the calculation, the median value of error is usually regarded as the location center, i.e.,

$$c(m) = \operatorname{median}\{e(m - N + 1), \ldots, e(m)\} \tag{8.15}$$

The sampling error median value is chosen as c because the presence of non-Gaussian impulsive noise in the error signal $e(m)$ will lead to large values of c when the sampling error mean value is used as c, thus reducing the robustness of the algorithm. Based on the above derivation, the pseudo code for KGMCC-VC is described in Algorithm 8.2, in which $\boldsymbol{\alpha}_i(m)$ and $\boldsymbol{\alpha}(m)$ denote the ith component of coefficient vector and the coefficient vector at iteration m.

8.4 Quantized Kernel Generalized Maximum Correntropy with Variable Centers

The KGMCC-VC algorithm iteratively updates the coefficient vector by training the sequential dataset $\mathcal{S}$ and it can be regarded as a growing RBF network. As a result, the computational complexity and memory demand increase as training data are continuously presented, which makes it difficult to apply the algorithm to applications with high real-time requirement. Since the sparsification method can effectively suppress the continuous growth of the network model, this subsection introduces the online vector quantization (VQ) method into the KGMCC-VC algorithm, which is called the QKGMCC-VC algorithm.

In brief, the QKGMCC-VC algorithm quantizes the feature vector $\boldsymbol{\psi}(\boldsymbol{x}(m))$. Thus, its weight update equation is expressed as

$$\boldsymbol{\omega}(m+1) = \boldsymbol{\omega}(m) + \eta g(e(m))\mathcal{Q}\big[\boldsymbol{\psi}(\boldsymbol{x}(m))\big] \tag{8.16}$$

where $\mathcal{Q}[\cdot]$ is quantization operator in RKHS. In fact, this quantization method cannot be directly applied in the KGMCC-VC algorithm due to the high dimensionality of the feature vector. In addition, for a Gaussian kernel, there was a positive correlation between the distance of the original input space and the distance of the feature space [24]. Therefore, the quantization operation is usually performed in the original space, i.e.,

$$\boldsymbol{\omega}(m+1) = \boldsymbol{\omega}(m) + \eta g(e(m))k_\sigma(Q[\boldsymbol{x}(m)], \cdot) \tag{8.17}$$

where $Q[\cdot]$ denotes quantization operator in the original space.

The online VQ method usually selects the appropriate codebook vector (dictionary) by the Euclidean metric of the input space, that is

$$\mathrm{dis}(\boldsymbol{x}(i), \boldsymbol{x}(j)) = \|\boldsymbol{x}(i) - \boldsymbol{x}(j)\| \tag{8.18}$$

When the distance between the input data and the dictionary is less than the quantization size ρ, the dictionary remains unchanged and, distinct from the traditional sparsification methods, updates locally the coefficients closest to the center using "redundant" data. Otherwise, the input data is appended to the dictionary and the corresponding coefficients are assigned. Hence, the proposed QKGMCC-VC is summarized in Algorithm 8.3.

8.5 Performance Analysis

The unknown system is considered as a nonlinear regression model, similar to other kernel adaptive filtering algorithms, i.e.,

$$d(m) = \boldsymbol{\omega}_{\mathrm{opt}}^{\mathrm{T}} \boldsymbol{\psi}(\boldsymbol{x}(m)) + v(m) \tag{8.19}$$

where $\boldsymbol{\omega}_{\mathrm{opt}}$ is the system weight vector in feature space $\mathbb{F}$, and $v(m)$ is the additive noise. To derive the energy conservation relation in feature space $\mathbb{F}$, the a priori error and the a posteriori error are defined as

$$e_a(m) = \widetilde{\boldsymbol{\omega}}(m)^{\mathrm{T}} \boldsymbol{\psi}(\boldsymbol{x}(m)) \tag{8.20}$$

$$e_p(m) = \widetilde{\boldsymbol{\omega}}(m+1)^{\mathrm{T}} \boldsymbol{\psi}(\boldsymbol{x}(m)) \tag{8.21}$$

where $\widetilde{\boldsymbol{\omega}}(m) = \boldsymbol{\omega}_{\mathrm{opt}} - \boldsymbol{\omega}(m)$ is weight error vector. Subtracting $\boldsymbol{\omega}_{\mathrm{opt}}$ from both sides of the weight update Eq. (8.12), we obtain

$$\widetilde{\boldsymbol{\omega}}(m+1) - \widetilde{\boldsymbol{\omega}}(m) = \eta g(e(m)) \boldsymbol{\psi}(\boldsymbol{x}(m)) \tag{8.22}$$

Note that the a priori error and the a posteriori error have the following relationship:

$$e_p(m) = e_a(m) + (\widetilde{\boldsymbol{\omega}}(m+1) - \widetilde{\boldsymbol{\omega}}(m))^{\mathrm{T}} \boldsymbol{\psi}(\boldsymbol{x}(m)) \tag{8.23}$$

Substituting (8.22) into (8.23), the a posteriori error becomes

$$e_p(m) = e_a(m) - \eta g(e(m)) k_\sigma(\boldsymbol{x}(m), \boldsymbol{x}(m)) \tag{8.24}$$

Finally, combining (8.22) and (8.24) to eliminate $g(e(m))$ yields

$$\widetilde{\boldsymbol{\omega}}(m+1) = \widetilde{\boldsymbol{\omega}}(m) + \frac{\big(e_p(m) - e_a(m)\big) \boldsymbol{\psi}(\boldsymbol{x}(m))}{k_\sigma(\boldsymbol{x}(m), \boldsymbol{x}(m))} \tag{8.25}$$

Squaring and taking expectation operations both sides of (8.25), one can obtain the following energy conservation relation

$$E\big[\|\widetilde{\boldsymbol{\omega}}(m+1)\|_{\mathbb{F}}^2\big] + E\left[\frac{e_a^2(m)}{k_\sigma(\boldsymbol{x}(m), \boldsymbol{x}(m))}\right] = E\big[\|\widetilde{\boldsymbol{\omega}}(m)\|_{\mathbb{F}}^2\big] + E\left[\frac{e_p^2(m)}{k_\sigma(\boldsymbol{x}(m), \boldsymbol{x}(m))}\right] \tag{8.26}$$

where $E\big[\|\widetilde{\boldsymbol{\omega}}(m)\|_{\mathbb{F}}^2\big]$ is the weight error power (WEP) in feature space $\mathbb{F}$.

Remark 8.4 When the kernel function is selected as Gaussian kernel, $k_\sigma(\boldsymbol{x}(m), \boldsymbol{x}(m)) = 1$ holds. In addition, the energy conservation relationship shown above is similar in form to the traditional linear adaptive filtering. However, their meanings are not completely the same. The energy conservation relation of KGMCC-VC algorithm exposes the evolution of WEP in the feature space.

Substituting (8.24) into (8.26) yields

$$E\left[\|\widetilde{\boldsymbol{\omega}}(m+1)\|_{\mathbb{F}}^2\right] = E\left[\|\widetilde{\boldsymbol{\omega}}(m)\|_{\mathbb{F}}^2\right] + \eta^2 E\left[g^2(e(m))\right] - 2\eta E\left[e_a(m)g(e(m))\right] \quad (8.27)$$

With the above conclusion, we can proceed to the subsequent derivation of the convergence condition as well as the steady-state behavior.

8.5.1 Stability Analysis

From (8.27), for any initial condition, the KGMCC-VC algorithm (8.12) converges asymptotically if the step size satisfies

$$\eta \leq \frac{2E\left[e_a g(e(m))\right]}{E\left[g^2(e(m))\right]} \frac{2E\left[e_a \exp\left(-\lambda|\in(m)|^{\zeta}\right)|\in(m)|^{\zeta-1}\mathrm{sign}(\in(m))\right]}{E\left[\exp\left(-2\lambda|\in(m)|^{\zeta}\right)|\in(m)|^{2(\zeta-1)}\right]} \quad (8.28)$$

where $\in(m) = e(m) - c$. Unfortunately, the bounds of (8.28) are not easily available in practical applications. In order to provide more details about the stability of the KGMCC-VC algorithm, the probability of divergence (POD) is examined. When no noise is involved as in [31], the desired signal can be expressed as $y(m) = \boldsymbol{\omega}_{\text{opt}}\boldsymbol{\psi}(\boldsymbol{x}(m))$. Thus, the weight error update equation is rewritten as

$$\begin{aligned}\widetilde{\boldsymbol{\omega}}(m+1) &= \widetilde{\boldsymbol{\omega}}(m) - \eta g(e(m))\boldsymbol{\psi}(\boldsymbol{x}(m))\left[1 - \eta\exp\left(-\lambda|\in(m)|^{\zeta}\right)|\in(m)|^{\zeta-2}\right]\widetilde{\boldsymbol{\omega}}(m) \\ &= \left[1 - \eta\exp\left(-\lambda|\in(m)|^{\zeta}\right)|\in(m)|^{\zeta}\|\widetilde{\boldsymbol{\omega}}(m)\|_{\mathbb{F}}^{-2}\right]\widetilde{\boldsymbol{\omega}}(m) \end{aligned} \quad (8.29)$$

where

$$0 \leq \exp\left(-\lambda|\in(m)|^{\zeta}\right)|\in(m)|^{\zeta} \leq \frac{1}{\lambda}\exp(-1)$$

Therefore, if $\|\widetilde{\boldsymbol{\omega}}(m)\|_{\mathbb{F}}^2 \geq \frac{\eta}{2\lambda}\exp(-1)$, the following inequation holds.

$$0 \leq \eta\exp\left(-\lambda|\in(m)|^{\zeta}\right)|\in(m)|^{\zeta}\|\widetilde{\boldsymbol{\omega}}(m)\|_{\mathbb{F}}^{-2} \leq 2 \quad (8.30)$$

It means

$$\left|1 - \eta\exp\left(-\lambda|\in(m)|^{\zeta}\right)|\in(m)|^{\zeta}\|\widetilde{\boldsymbol{\omega}}(m)\|_{\mathbb{F}}^{2}\right| \leq 1 \quad (8.31)$$

Substituting (8.31) into (8.29) yields

$$\|\widetilde{\boldsymbol{\omega}}(m+1)\|_{\mathbb{F}}^2 \le \|\widetilde{\boldsymbol{\omega}}(m)\|_{\mathbb{F}}^2 \tag{8.32}$$

Remark 8.5 From the above derivation, we can see that $\lim_{m\to\infty} |\boldsymbol{\omega}(m)| < \infty$ holds if the bound of $|\boldsymbol{\omega}(m)|$ exists, which means that the POD of the KGMCC-VC algorithm is zero in the absence of noise. This proves the robustness of the proposed algorithm in a sense. Unfortunately, it is difficult to verify the conclusion in the simulation section because the optimal weights $\boldsymbol{\omega}_{\mathrm{opt}}$ are not available in the kernel adaptive filtering.

8.5.2 Steady-State Mean Square Performance

To analyze the steady-state behavior of the KGMCC-VC algorithm, the steady-state excess mean square error (EMSE) is defined as

$$\tau_{\mathrm{EMSE}} = \lim_{m\to\infty} E\left[e_a^2(m)\right] \tag{8.33}$$

Before proceeding to the subsequent derivation, the following two common assumptions are given:

A1 The measurement noise $v(m)$ is independent, identically distributed (i.i.d.) process with zero mean.
A1 The a priori error $e_a(m)$ is independent of the measurement noise $v(m)$, and $E[e_a(m)] = 0$.

Suppose that the filter is stable and reaches a steady state after a while, so that we have

$$\lim_{m\to\infty} \|\widetilde{\boldsymbol{\omega}}(m+1)\|_{\mathbb{F}}^2 = \lim_{m\to\infty} \|\widetilde{\boldsymbol{\omega}}(m)\|_{\mathbb{F}}^2 \tag{8.34}$$

Combining (8.27) and (8.34) leads to

$$2E\left[e_a(m)g(e(m))\right] = \eta E\left[g^2(e(m))\right] \tag{8.35}$$

The distribution of $e_a(m)$ and $e(m)$ in the steady state is independent of m. Therefore, it can be abbreviated as

$$2E\left[e_a g(e)\right] = \eta E\left[g^2(e)\right] \tag{8.36}$$

By Taylor Expansion [19], $g(e)$ is expressed as

$$g(e) = g(v + e_a) = g(v) + g'(v)e_a + \frac{1}{2}g''(v)e_a^2 + o\left(e_a^2\right) \tag{8.37}$$

where $g'(v)$ and $g''(v)$ denote the first-order and second-order derivatives, $o(e_a^2)$ is the third and higher-order terms, and $g'(v)$ and $g''(v)$ is written as

$$g'(v) = \exp\left(-\lambda|v-c|^{\zeta}\right)|v-c|^{\zeta-2}\left(\zeta-1-\lambda\zeta|v-c|^{\zeta}\right) \tag{8.38}$$

$$\begin{aligned} g''(v) = &\exp\left(-\lambda|v-c|^{\zeta}\right)|v-c|^{\zeta-3} \\ &\left((\zeta-1)(\zeta-2)-3\lambda\zeta(\zeta-1)|v-c|^{\zeta}+\lambda^2\zeta^2|v-c|^{2\zeta}\right) \end{aligned} \tag{8.39}$$

With assumptions A1 and A2, if $o(e_a^2)$ is small enough, then it can be derived

$$E\left[e_a g(e)\right] = E\left[e_a g(v) + g'(v)e_a^2 + o(e_a^2)\right] \approx E\left[g'(v)\right]\tau_{\text{EMSE}} \tag{8.40}$$

$$E\left[g^2(e)\right] \approx E\left[g^2(v)\right] + E\left[g(v)g''(v) + |g'(v)|^2\right]\tau_{\text{EMSE}} \tag{8.41}$$

Substituting (8.40) and (8.41) into (8.36) yields

$$\tau_{\text{EMSE}} = \frac{\eta E\left[g^2(e)\right]}{2E\left[g'(v)\right] - \eta E\left[g(v)g''(v) + |g'(v)|^2\right]} \tag{8.42}$$

According to (8.38), (8.39), and (8.42), one can obtain

$$\tau_{\text{EMSE}} = \frac{\eta E\left[\exp\left(-2\lambda|v-c|^{\zeta}|v-c|^{2\zeta-2}\right)\right]}{2E\left[\exp\left(-\lambda|v-c|^{\zeta}\right)|v-c|^{\zeta-2}\left(\zeta-1-\lambda\zeta|v-c|^{\zeta}\right)\right] - \eta E[\varkappa(v)]} \tag{8.43}$$

where

$$\begin{aligned} \varkappa(v) = &\exp\left(-2\lambda|v-c|^{\zeta}\right)|v-c|^{2\zeta-4} \\ &\left((\zeta-1)(2\zeta-3)-5\lambda\zeta(\zeta-1)|v-c|^{\zeta}+2\lambda^2\zeta^2|v-c|^{2\zeta}\right) \end{aligned} \tag{8.44}$$

If the step-size is small enough, (8.43) is written as

$$\tau_{\text{EMSE}} \approx \frac{\eta E\left[\exp\left(-2\lambda|\epsilon|^{\zeta}|\epsilon|^{2\zeta-2}\right)\right]}{2E\left[\exp\left(-\lambda|\epsilon|^{\zeta}\right)|\epsilon|^{\zeta-2}\left(\zeta-1-\lambda\zeta|\epsilon|^{\zeta}\right)\right]} \tag{8.45}$$

where $\epsilon = v - c$.

Remark 8.6 Note that the above steady-state performance analysis is concluded when the higher-order terms of a priori error and step sizes are small enough. When they cannot be ignored, the simulation results will not match well with the theoretical values.

8.6 Simulation Results

In this section, the simulation results of nonlinear system modeling and MG time series prediction are provided to validate the theoretical model and the performance of the KGMCC-VC algorithm, respectively.

8.6.1 *Steady-State Performance*

First, we examine the correctness of the proposed theoretical model. The desired signal is generated by the following nonlinear system [28]:

$$d(m) = a_0 \sin(x(m)) + a_1 x(m-1) + a_2 x^2(m-2) + v(m) \tag{8.46}$$

where the parameter vector $[a_0, a_1, a_2]^{\mathrm{T}} = [1, 0.5, -0.1]^{\mathrm{T}}$, the input data $\{x(m)\}$ is obtained by a white Gaussian process with unit variance. In the subsequent simulations, we take $x(m) = [x(m-2), x(m-1), x(m)]^{\mathrm{T}}$ as the input vector and $d(m)$ as the output of the system with noise.

In the simulation, the shape parameter ζ is 4, the kernel parameter λ is 2, and the kernel bandwidth σ is 1. In addition, we choose a uniform distribution over $[0,\ b]$ as the noise. In order to ensure that the KGMCC-VC algorithm reaches steady state, 200,000 iterations are run. The simulated EMSEs is obtained by averaging the last 1000 iterations. Figures 8.1, 8.2 and 8.3 show the simulated and theoretical EMSEs for different values of step size, noise variance, and center, respectively. It can be seen from the above figures that the EMSEs of the theoretical model matches well with the EMSEs of the simulation when the noise variance and step size are small, and the center location c is close to the noise mean. On the contrary, the theoretical model does not predict the EMSE behavior of the KGMCC-VC algorithm well. In addition, the steady-state EMSE increases with step-size η, noise variance, and distance between the noise center and center location of the proposed algorithm.

8.6.2 *Performance Comparison*

Second, we compare the proposed KGMCC-VC algorithm with other kernel adaptive filtering algorithms in the scenario where the desired signal contains different types of noise. In the simulation, the kernel size σ, and the kernel parameter λ are set as 1, and 2. The center location c is simply set to the mean value of the corresponding noise. The step size is chosen so that all algorithms have almost the same initial convergence rate. In addition, four noises with different distribution are considered: (1) Gaussian distribution with mean 1 and variance 0.25; (2) Uniform distribution

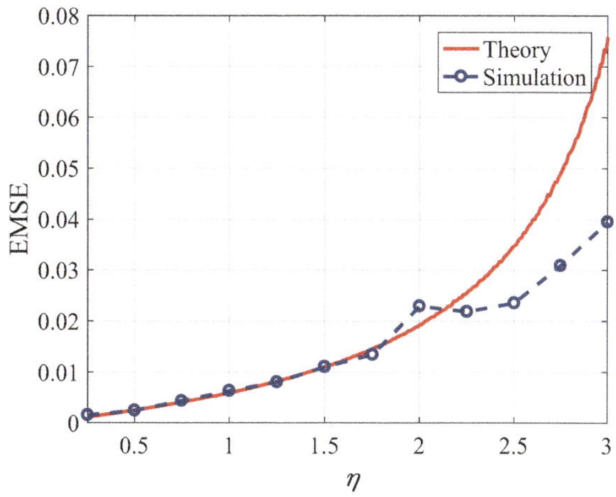

Fig. 8.1 Simulated and theoretical EMSEs of KGMC-VC algorithm with different step size η ($b = 0.5,\ c = 0.25$)

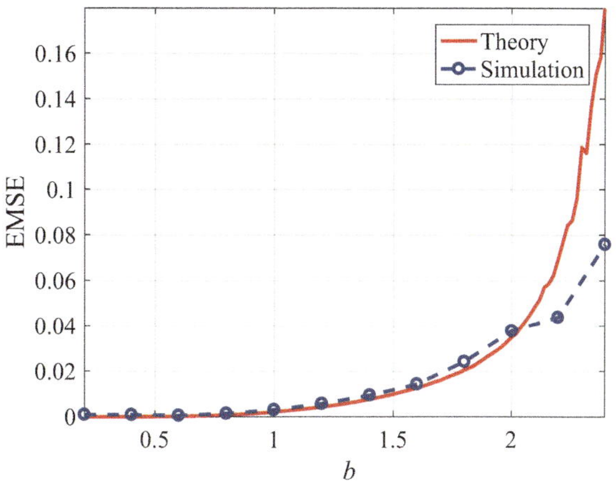

Fig. 8.2 Simulated and theoretical EMSEs of KGMC-VC algorithm with different noise variances ($\eta = 0.5,\ c = 0.25$)

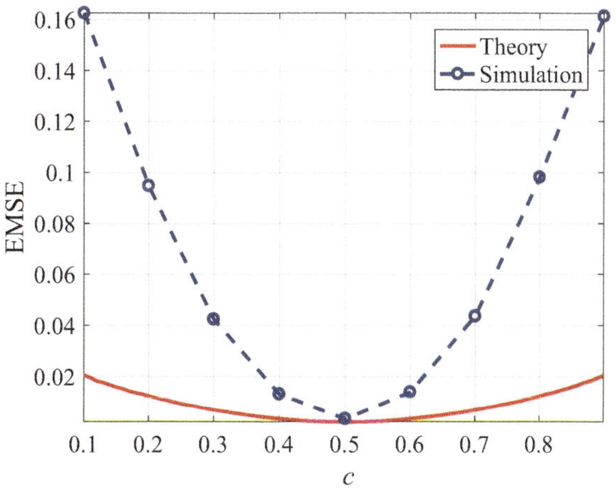

Fig. 8.3 Simulated and theoretical EMSEs of KGMC-VC algorithm with different center location c ($\eta = 0.5,\ b = 1$)

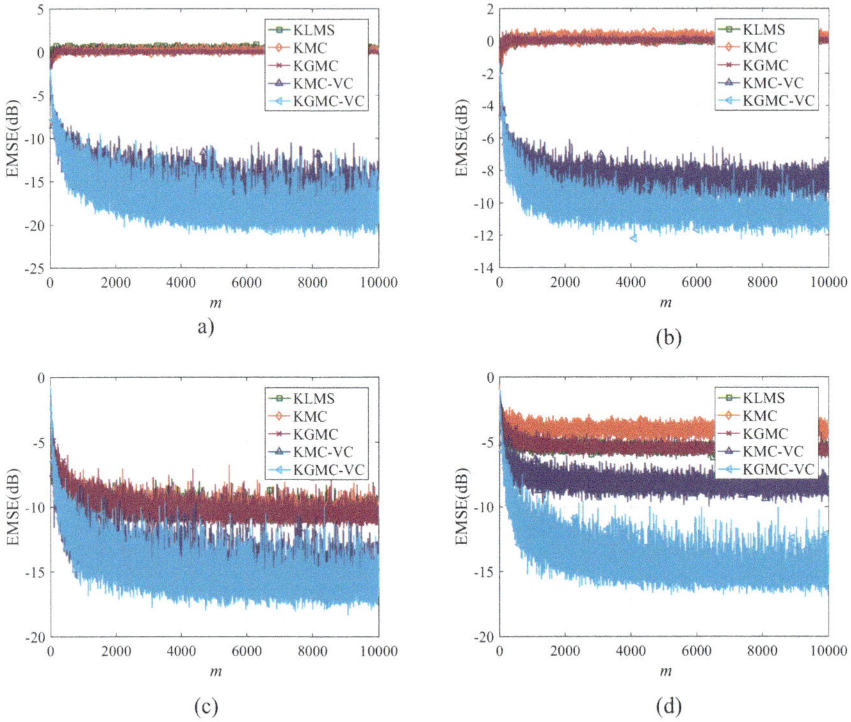

Fig. 8.4 Learning curves for the EMSE with different noise distributions: **a** Gaussian; **b** Uniform; **c** Laplace; **d** Binary

over [0 2]; (3) Laplace distribution with mean 0.25 and variance 0.25; (4) Binary distribution over $\{0,\ 1\}$ with probability mass $p[v=0]=p[v=1]=0.5$.

The EMSE convergence curves are obtained over 200 independent Monte Carlo runs. Figure 8.4a–d shows the EMSE convergence curves of KGMCC-VC and other kernel adaptive filtering algorithms under different noise environments, respectively. As expected, the KGMCC-VC algorithm has excellent performance in all four noise environments compared with other kernel adaptive filtering algorithms. Especially, when the noise mean is far from zero, the performance of KLMS, KMCC, and KGMCC algorithms will deteriorate significantly as shown in Fig. 8.4a and b.

8.6.3 Performance of QKGMCC-VC

Third, we compare the network size and convergence performance of the KGMCC-VC and QKGMCC-VC algorithms. In the simulation, the kernel size σ, the step-size η, the center location c, and the kernel parameter λ are set as 1, 1.2, 0.5, and 2. The noise is chosen to be binary distributed over $\{0,\ 1\}$ with probability mass $p[v=0]=$

$p[v = 1] = 0.5$. The quantization size ρ is fixed to 0.5. The EMSE learning curve and network size with respect to iteration for the KGMCC-VC algorithm and QKGMCC-VC algorithm are given in Figs. 8.5 and 8.6, respectively. It can be seen by combining Figs. 8.5 and 8.6 that the dictionary dimension of the QKGMCC-VC algorithm is significantly lower than that of the KGMCC-VC algorithm, while their EMSEs are almost the same. This indicates that the sparsification method (quantization) can effectively reduce the computational complexity while maintaining the accuracy of the algorithm.

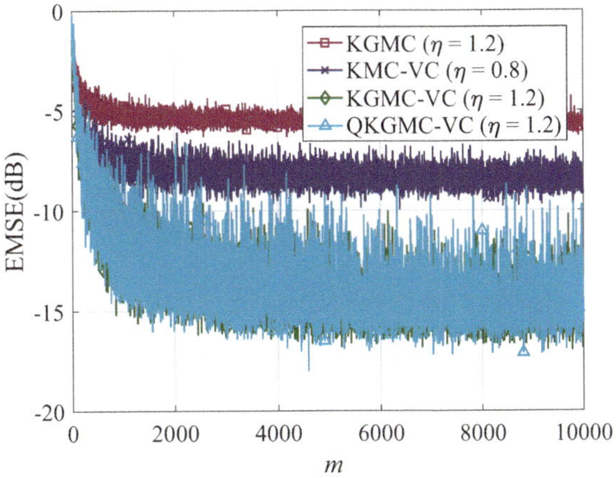

Fig. 8.5 EMSE convergence curves of KGMC-VC and QKGMC-VC algorithms

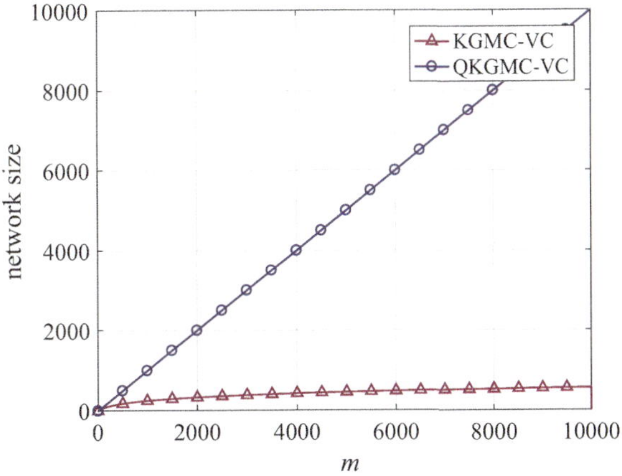

Fig. 8.6 Network size growth of KGMC-VC and QKGMC-VC algorithms

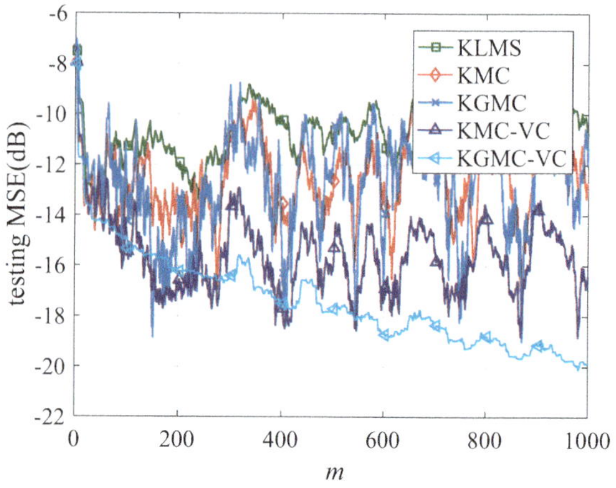

Fig. 8.7 Testing MSE convergence curves of the different kernel adaptive filtering algorithms

8.6.4 *Application to Time Series Prediction*

Finally, we consider the Mackey–Glass chaotic time series prediction, which is generated from the following differential equation.

$$\frac{\mathrm{d}x(t)}{\mathrm{d}t} = -a_0 x(t) + \frac{a_1 x(t-\tau)}{1 + x(t-\tau)^{10}} \tag{8.46}$$

where $[a_0, a_1, \tau]^{\mathrm{T}} = [0.1, 0.2, 30]^{\mathrm{T}}$. A total of 6000 observation sequences is generated with a sampling time of 6s [10]. The time embedding is 10, and 1000 samples are used as the training data and another 100 as the test data. All data are disturbed by the uniform distribution noise over [0, 0.4]. In the simulation, we compare the testing MSEs of five kernel adaptive filtering algorithms, where the kernel size σ, the kernel parameter λ, and the sample number N are fixed to 1, 1, and 20. The step size is chosen to ensure that the different algorithms have the same initial convergence rate. The testing MSE convergence curves of the different kernel adaptive filtering algorithms are given in Fig. 8.7, and the simulation results show that the proposed algorithm has the lowest test MSE value in the nonzero-mean noise environment.

8.7 Summary

In this paper, we propose a kernel adaptive filtering algorithm of KGMCC-VC and its quantized version. Compared with existing kernel adaptive filtering algorithms, the proposed algorithm has lower steady-state performance, especially in the case of nonzero-mean noise. In addition, the computational complexity of the KGMCC-VC and QKGMCC-VC is similar to that of KLMS and QKLMS. Based on the energy conservation relation, we also provide the mean-square steady-state behavior and convergence condition of the KGMCC-VC algorithm. The simulation results validate the superiority of the proposed algorithm. Unfortunately, when the a priori error is relatively large or the step size is relatively large, discrepancies between the theoretical model and the Monte Carlo simulation will occur. How to overcome this challenge is the subject of future research.

References

1. S. Ertekin, L. Bottou, C.L. Giles, Nonconvex online support vector machines. IEEE Trans. Pattern Anal. Mach. Intell. **33**, 368–381 (2011)
2. W. Zhang, H. Zhang, J. Liu, K. Li, D. Yang, T. Hui, Weather prediction with multiclass support vector machines in the fault detection of photovoltaic system. IEEE/CAA J. Automatica Sinica **4**, 543–550 (2017)
3. T.J. Chin, D. Suter, Incremental kernel principal component analysis. IEEE Trans. Image Process. **16**, 1662–1674 (2007)
4. J. C. Príncipe, W. Liu, S. Haykin, *Kernel Adaptive Filtering: A Comprehensive Introduction* (Wiley, 2011)
5. C. Chen, W. Li, H. Su, K. Liu, Spectral-spatial classification of hyperspectral image based on kernel extreme learning machine. Remote sensing **6**, 5795–5814 (2014)
6. S. An, W. Liu, S. Venkatesh, Face recognition using kernel ridge regression, in *2007 IEEE Conference on Computer Vision and Pattern Recognition* (2007), pp. 1–7
7. L. Tan, J. Jiang, Adaptive volterra filters for active control of nonlinear noise processes. IEEE Trans. Signal Process. **49**, 1667–1676 (2001)
8. L. Lu, H. Zhao, B. Chen, Time series prediction using kernel adaptive filter with least mean absolute third loss function. Nonlinear Dyn. **90**, 999–1013 (2017)
9. W. Liu, I. Park, J.C. Principe, An information theoretic approach of designing sparse kernel adaptive filters. IEEE Trans. Neural Networks **20**, 1950–1961 (2009)
10. W. Liu, P.P. Pokharel, J.C. Principe, The kernel least-mean-square algorithm. IEEE Trans. Signal Process. **56**, 543–554 (2008)
11. T.K. Paul, T. Ogunfunmi, Study of the convergence behavior of the complex kernel least mean square algorithm. IEEE Trans. Neural Netw. Learn. Syst. **24**, 1349–1363 (2013)
12. T.K. Paul, T. Ogunfunmi, A kernel adaptive algorithm for quaternion-valued inputs. IEEE Trans. Neural Netw. Learn. Syst. **26**, 2422–2439 (2015)
13. W. Liu, J.C. Príncipe, Kernel affine projection algorithms. EURASIP J. Adv. Signal Process. **2008**, 1–12 (2008)
14. Y. Engel, S. Mannor, R. Meir, The kernel recursive least-squares algorithm. IEEE Trans. Signal Process. **52**, 2275–2285 (2004)
15. S.V. Vaerenbergh, M. Lazaro-Gredilla, I. Santamaria, Kernel recursive least-squares tracker for time-varying regression. IEEE Trans. Neural Netw. Learn. Syst. **23**, 1313–1326 (2012)
16. S. Zhao, B. Chen, J. C. Príncipe, Kernel adaptive filtering with maximum correntropy criterion, in *The 2011 International Joint Conference on Neural Networks* (2011), pp. 2012–2017
17. B. Chen, Z. Yuan, N. Zheng, J.C. Príncipe, Kernel minimum error entropy algorithm. Neurocomputing **121**, 160–169 (2013)
18. B. Shen, X. Wang, L. Zou, Maximum correntropy kalman filtering for non-Gaussian systems with state saturations and stochastic nonlinearities. IEEE/CAA J. Automatica Sinica **10**, 1–11 (2023)
19. B. Chen, L. Xing, H. Zhao, N. Zheng, J.C. Principe, Generalized correntropy for robust adaptive filtering. IEEE Trans. Signal Process. **64**, 3376–3387 (2016)
20. Y. Zhu, H. Zhao, X. Zeng, B. Chen, Robust generalized maximum correntropy criterion algorithms for active noise control. IEEE/ACM Trans. Audio Speech Lang. Process. **28**, 1282–1292 (2020)

21. J. He, G. Wang, B. Peng, Z. Feng, K. Zhang, Generalized minimum error entropy for adaptive filtering. Pattern Recogn. **135**, 109188 (2023)
22. X. Liu, C. Song, Z. Pang, Kernel recursive maximum correntropy with variable center. Signal Process. **191**, 108364 (2022)
23. Q. Sun, H. Zhang, X. Wang, W. Ma, B. Chen, Sparsity constrained recursive generalized maximum correntropy criterion with variable center algorithm. IEEE Trans. Circ. Syst. II Express Briefs **67**, 3517–3521 (2020)
24. B. Chen, S. Zhao, P. Zhu, J.C. Principe, Quantized kernel least mean square algorithm. IEEE Trans. Neural Netw. Learn. Syst. **23**, 22–32 (2012)
25. W.D. Parreir, J.C.M. Bermudez, C. Richard, J.Y. Tourneret, Stochastic behavior analysis of the Gaussian kernel least-mean-square algorithm. IEEE Trans. Signal Process. **60**, 2208–2222 (2012)
26. W.D. Parreira, M.H. Costa, J.C. Bermudez, Stochastic behavior analysis of the Gaussian KLMS algorithm for a correlated input signal. Signal Process. **152**, 286–291 (2018)
27. V.R. Elias, V.C. Gogineni, W.A. Martins, S. Werner, Adaptive graph filters in reproducing kernel Hilbert spaces: design and performance analysis. IEEE Trans. Signal Inf. Process. Netw. **7**, 62–74 (2020)
28. B. Chen, S. Zhao, P. Zhu, J.C. Príncipe, Mean square convergence analysis for kernel least mean square algorithm. Signal Process. **92**, 2624–2632 (2012)
29. W. Liu, P.P. Pokharel, J.C. Principe, Correntropy: Properties and applications in non-Gaussian signal processing. IEEE Trans. Signal Process. **55**, 5286–5298 (2007)
30. Y. He, F. Wang, J. Yang, H. Rong, B. Chen, Kernel adaptive filtering under generalized maximum correntropy criterion, in *2016 International Joint Conference on Neural Networks (IJCNN)* (2016), pp. 1738–1745
31. V. Nascimento, J. Bermudez, Probability of divergence for the least-mean fourth algorithm. IEEE Trans. Signal Process. **54**, 1376–1385 (2006)

Chapter 9
Diffusion Kernel Generalized Maximum Correntropy with Variable Centers

The existing adaptive filtering algorithms in distributed networks do not take into account the case of nonzero-mean non-Gaussian noise. In response to this issue, a robust diffusion kernel adaptive filtering algorithm, which is called the diffusion kernel generalized maximum correntropy criterion with variable center (DKGMCC-VC) algorithm, is proposed. The algorithm aims to address a noisy environment with nonzero mean by dynamically adjusting the center of the kernel function in the generalized correntropy. An online vector quantization method is also introduced, which can be used to reduce the complexity of the algorithm. In addition, the convergence condition and steady-state performance of the algorithm are analyzed. Finally, simulation results verify the robustness and superiority of the algorithm in dealing with non-Gaussian noise environments with nonzero mean in distributed models.

9.1 Introduction

Adaptive filtering algorithms have long served as a cornerstone in signal processing, owing to their powerful self-learning and self-tracking capabilities. Traditional approaches exemplified by least mean square (LMS) [1] and recursive least squares (RLS) [2] algorithms demonstrate inherent limitations when confronting nonlinear system modeling challenges. This critical gap motivated the emergence of kernel adaptive filtering (KAF) methodologies, which synergistically integrate kernel methods with conventional adaptive filtering frameworks. This method effectively handles nonlinear system modeling problems by constructing the kernel function, which maps the input data to the reproducing kernel Hilbert space (RKHS) and then uses the linear algorithms to process them. This has produced several influential algorithms including kernel least mean square (KLMS) [3–5], kernel recursive least

H. Zhao et al., *Information Theoretic Learning-Based Filter*,
https://doi.org/10.1007/978-3-032-29623-8_9

squares (KRLS) [6, 7], kernel affine projection (KAP) [8], and extended KRLS (Ex-KRLS) [9]. They have demonstrated particular efficacy in meteorological time-series prediction tasks encompassing temperature, humidity, and wind speed forecasting.

Although there has been some progress, conventional KAF architectures exhibit diminished robustness in non-Gaussian noise environments where standard minimum mean square error (MMSE) is not very effective. This limitation has promoted the development of information-theoretic learning frameworks [10, 11], and developed the corresponding KAF algorithms such as the kernel maximum correntropy criterion (KMCC) [12] and kernel minimum error entropy (KMEE) [13] algorithms. They involve higher-order error information, so compared to MMSE-based approaches, these information-theoretic variants demonstrate superior performance in impulsive noise scenarios through their inherent capacity to exploit higher-order error statistics [14, 15], thereby new directions are provided for adaptive filtering in non-Gaussian environments.

In recent years, big data has posed a significant challenge to traditional single-node learning methods. How to achieve high-precision parameter estimation in distributed networks has become a new research hotspot [16]. The basic concept of distributed networks is to spread computational tasks among multiple network nodes, making full use of each node's [17] computational capacity and localized information. Distributed learning has applications in various fields such as spectrum sensing [18], frequency estimation in power networks [19], active noise control [20, 21], and so on. In the field of distributed estimation, diffusion adaptation strategies are attracting much attention for their efficiency and scalability. In previous work, the scholars have proposed distributed algorithms based on diffusion strategies, such as diffusion recursive least squares (DRLS) [22, 23], proportionate robust diffusion recursive least exponential hyperbolic cosine (PR-DREHC) [24], diffusion least mean square (DLMS) [25–27], minimum disturbance diffusion least mean square (MD-DLMS) [28] and diffusion sparse least mean square (DSLMS) [29]. These algorithms enabling global parameter estimation through localized node cooperation, and can improve filtering results by fusing information from nodes within the same neighbor, these strategies achieve accelerated convergence rates and reduced steady-state mean square error (MSE) compared to a single node [30]. This promotes the use of distributed algorithms in different fields, including power spectrum analysis, system identification, and cognitive radio.

However, a key limitation remains: traditional distributed diffusion algorithms rely heavily on MMSE-based cost functions and are prone to performance degradation in non-Gaussian noise environments. Therefore, in recent years, in order to improve the performance of algorithms in non-Gaussian environments of distributed networks, scholars have proposed several robust distributed adaptive filtering algorithms, such as the diffusion maximum correntropy criterion (DMCC) [31] algorithm and so on. Meanwhile. it is very critical to choose an appropriate kernel in the correntropy, in which the Gaussian kernel function is widely used because of its simplicity and effectiveness, but the Gaussian kernel is not always optimal. In order to improve the performance of the DMCC algorithm, the generalized maximum correntropy criterion (GMCC) [32] was proposed by combining with the generalized Gaussian density (GGD) function. The diffusion generalized maximum correntropy criterion (DGMCC) [33] algorithm is proposed by combining it with distributed networks.

Unfortunately, when the noise mean is not zero, the performance of the above algorithms decreases significantly, thus leading to a substantial decrease in their effectiveness and accuracy, and failing to achieve the expected results, which may affect the overall reliability and robustness of these algorithms in practical applications. The MCC-based variable center algorithm proposed in [34] can effectively solve this problem. However, the above linear algorithm is not very effective in dealing with nonlinear modeling problems. In order to solve the problems, this paper proposes a diffusion kernel generalized maximum correntropy criterion with variable center (DKGMCC-VC) algorithm. In order to reduce the growth of computational complexity, an online vector quantization (VQ) [35] sparsification method is introduced and the quantized DKGMCC-VC (DQKGMCC-VC) algorithm is proposed.

The main contributions of our work are summarized as follows:

This article addresses the limitations of the existing MEE criterion-based CKF (MEE-CKF) for SOC estimation under complex non-Gaussian noise conditions. A new cost function is constructed by combining two MEE kernel functions with different kernel widths using a mixing factor. Additionally, both Gaussian kernels are replaced with a generalized Gaussian kernel. Moreover, the proposed algorithm constitutes a significant advantage of SRCKF in addressing numerical problems. The employment of mixing factor offers a more adaptable approach to address non-Gaussian noise. The main contributions are as follows:

(1) A diffusion kernel adaptive filtering algorithm with variable center, the DKGMCC-VC algorithm, is proposed, and a quantization version of the algorithm is presented, which meets the challenge of modeling nonlinear systems with nonzero-mean non-Gaussian noises in distributed models.
(2) Through theoretical analysis, the paper discusses the steady-state performance of DKGMCC-VC algorithm, which provides a solid theoretical basis for the performance evaluation of the algorithm in practical application.
(3) Based on the simulation study, this paper verifies the validity and superiority of the proposed algorithm in handling nonlinear signal problems in the background of nonzero-mean and non-Gaussian noise in a distributed system.

9.2 Diffusion Kernel Generalized Maximum Correntropy with Variable Centers Algorithm

In distribution, each network consists of N nodes, and each node communicates with other adjacent nodes alone at time m; then the output model of each node n can be represented as:

$$d_n(m) = \boldsymbol{\omega}_0^{\mathrm{T}} \varphi(\boldsymbol{x}_n(m)) + v_n(m), \tag{9.1}$$

v_n is the environment noise with nonzero-mean, where $\boldsymbol{\omega}_0 = [\omega_o, \omega_1, \ldots, \omega_{L-1}]^{\mathrm{T}}$ is the unknown weight vector to be estimated, $\varphi : \mathbb{R}^L \to \mathbb{F}$ is the feature map, and $\boldsymbol{x}_n(m) = [x_n(m), x_n(m-1), \ldots, x_n(m-L+1)]^{\mathrm{T}}$ denotes the input signal vector. The system output error signal is:

$$e_n(m) = d_n(m) - \boldsymbol{\omega}_0^{\mathrm{T}} \varphi(\boldsymbol{x}_n(m)). \tag{9.2}$$

Consider a supervised machine learning task where we aim to create a mapping function $\boldsymbol{f} : \mathbb{R}^L \to \mathbb{R}$ based on a dataset $\boldsymbol{S} = \{x_n(m), d_n(m)\}_{n=1}^N$, where $x_n(m) \in \mathbb{R}^L$ represents the input vector at discrete time m, and the desired output is given by $d_n(m) = f(x_n(m)) + v_n(m)$, where $v_n(m)$ is additive noise. When the mapping function $\boldsymbol{f}$ is nonlinear, indicating a nonlinear underlying system, traditional linear adaptive filtering algorithms may not yield satisfactory results. Kernel methods, due to their strong nonlinear mapping capabilities, allow us to construct the nonlinear function $\boldsymbol{f}$ using linear adaptive filtering techniques in a high-dimensional feature space $\mathbb{F}$. Furthermore, the mapping function $\boldsymbol{f}$ is typically represented by a weight vector in kernel adaptive algorithms. According to the representer theorem, the mapping function $\boldsymbol{f}$ (or weight vector ω) can be expressed as:

$$\boldsymbol{f} = \boldsymbol{\omega} = \sum_{m=1}^{n} \alpha_n(m) \varphi(\boldsymbol{x}_n(m)) = \sum_{m=1}^{n} \alpha_n(m) \kappa_\sigma(\boldsymbol{x}_n(m), \cdot), \tag{9.3}$$

where $\{\alpha_n(m)\}$ is the expansion coefficient, $\varphi : \mathbb{R}^L \to \mathbb{F}$ is the feature map, and $\kappa_\sigma(\cdot, \cdot)$ is the kernel function. The kernel function generally uses the Gaussian kernel, and the Gaussian kernel function is expressed as

$$\kappa_\sigma(\boldsymbol{x}_n(i), \boldsymbol{x}_n(j)) = \exp\left(-\frac{\|\boldsymbol{x}_n(i) - \boldsymbol{x}_n(j)\|^2}{2\sigma^2}\right). \tag{9.4}$$

It should be noted that the kernel function of correntropy has different meanings from the kernel function of kernel methods. The former emphasizes the similarity measure in RKHS, while the latter focuses on simplifying complex operations in high-dimensional feature space. In other words, the latter can be used by replacing the inner product of the sum of two feature vectors by simple operations, i.e.,

$$\kappa_\sigma(\boldsymbol{x}_n(i), \boldsymbol{x}_n(j)) = \varphi(\boldsymbol{x}_n(i))^{\mathrm{T}}\varphi(\boldsymbol{x}_n(j)). \tag{9.5}$$

For the DKGMCC-VC algorithm, a diffusion network consisting of N nodes was used, where all nodes jointly estimate the unknown weight vectors $\boldsymbol{\omega}(m)$. The global cost function is defined as follows:

$$\begin{aligned} J_{\text{DGKMCC-VC}}^{\text{global}}[\boldsymbol{\omega}(m)] &= \sum_{n=1}^{N} J_{\text{DGKMCC-VC}}^{\text{local}}[\boldsymbol{\omega}(m)] \\ &= \sum_{n=1}^{N} \left(\gamma_{\xi,\theta} - E\left[G_{\xi,\theta}(e_n(m) - o_n\right]\right) \\ &= \sum_{n=1}^{N} \left\{\gamma_{\xi,\theta}\left(1 - E\left[\exp(-\lambda|e_n(m) - o_n|^{\xi}\right]\right)\right\} \end{aligned} \tag{9.6}$$

where $J_{\text{DGKMCC-VC}}^{\text{local}}$ is the local cost function in the diffusion network, defined as follows:

$$J_{\text{DGKMCC-VC}}^{\text{local}} = \gamma_{\xi,\theta}\left(1 - E\left[\exp(-\lambda|e_n(m) - o|^{\xi}\right]\right) \tag{9.7}$$

where o can be considered as the mean (median) value of the last N sampling errors at time m, $e_n(m)$ is the prediction error of the node n at time m:

$$e_n(m) = d_n(m) - \boldsymbol{\omega}_n(m-1)^{\mathrm{T}}\varphi(\boldsymbol{x}_n(m)) \tag{9.8}$$

In distributed network estimation, there are two commonly used strategies including adapt-then-combine (ATC) and combine-then-adapt (CTA). The ATC strategy is usually preferred over the CTA strategy [36]. Similarly, the ATCDKGMCC algorithm tends to outperform the CTADKGMCC algorithm due to the fact that the ATC strategy efficiently uses data from nodes two hops apart in each iteration, whereas the CTA version uses data from nodes that are one hop apart [31]. In the following section, the stochastic gradient algorithm and the ATC diffusion strategy will be employed. By calculating the partial derivatives of Eq. (9.7), we can obtain:

Algorithm 9.1: Pseudo-code for DKGMCC-VC algorithm

Input: Kernel bandwidth σ; Step-size μ; Kernel parameter λ; Shape parameter ζ;

Output: Coefficient vector $\boldsymbol{\alpha}$

/* Initialization */

$m \leftarrow 1,\ n \leftarrow 1$

while $n \le N$ **do**

  $o_n(1) \leftarrow 0$

  $\alpha_n(1) \leftarrow 0$

  $n \leftarrow n+1$

end

$n \leftarrow 1$

/* Computation */

while $m \le M$ **do**

  while $n \le N$ **do**

    $\hat{d}_n(m) \leftarrow \sum_{i=1}^{m-1} \alpha_{n,i}(m-1)\kappa_\sigma(\boldsymbol{x}_n(i), \boldsymbol{x}_n(m))$

    $e_n(m) \leftarrow d_n(m) - \hat{d}_n(m)$

    $\Lambda_n(m) \leftarrow e_n(m) - o_n(m-1)$

    $g(e_n(m)) \leftarrow \exp(-\lambda|\Lambda_n(m)|^{\zeta})|\Lambda_n(m)|^{\zeta-1} sign(\Lambda_n(m))$

    $\hat{\alpha}_n(m) \leftarrow \{\hat{\alpha}_n(m-1), \mu g(e_n(m))\}$

    $n \leftarrow n+1$

    if $m < M$ **then**

      $o_n(m+1) \leftarrow 0$

    else

      $o_n(m+1) \leftarrow median\{e_n(m-N+1), \cdots, e_n(m)\}$

    end

  end

  $\alpha_n(m) \leftarrow \sum_{l \in N_n} a_{l,n}\hat{\alpha}_n(m)$

  $m \leftarrow m+1, n \leftarrow 1$

end

return $\boldsymbol{\alpha}$

$$\begin{aligned}\nabla J^{\text{local}}_{\text{DGKMCC-VC}}[\boldsymbol{\omega}(m)] &= \frac{\partial}{\partial \boldsymbol{\omega}(m)}\left(\nabla J^{\text{local}}_{\text{DGKMCC-VC}}[\boldsymbol{\omega}(m)]\right)\\ &= \frac{\partial}{\partial \boldsymbol{\omega}(m)}\left(\gamma_{\xi,\theta}\left(1 - E\left[\exp\left(-\lambda|e_n(m) - o|^{\xi}\right)\right]\right)\right)\\ &= \mu g(e_n(m))\varphi(\boldsymbol{x}_n(m))\end{aligned} \tag{9.9}$$

where $\mu = \eta\lambda\xi$ is the step-size parameter, and

$$g(e_n(m)) = \exp\left(-\lambda|e_n(m) - o|^{\xi}\right)|e_n(m) - o|^{\xi-1}\text{sign}(e_n(m) - o) \tag{9.10}$$

The intermediate estimate of node n at moment m is

$$\boldsymbol{\phi}_n(m+1) = \boldsymbol{\omega}_n(m) + \mu g(e_n(m))\varphi(\boldsymbol{x}_n(m)) \tag{9.11}$$

where μ is the step size, $\boldsymbol{\omega}_n(m)$ is the estimate $\boldsymbol{\omega}_0$ of node n at time m. The result of the intermediate estimates for all adjacent nodes via node n is

$$\boldsymbol{\omega}_n(m+1) = \sum_{l\in N_n} a_{l,n}\boldsymbol{\phi}_l(m+1) \tag{9.12}$$

where N_n denotes the set of neighboring nodes of node n.

Remark 9.1 It can be seen that the above weight update equation degenerate to different diffusion kernel adaptive filtering algorithms: (1) the DKMCC-VC algorithm when $\xi = 2$; (2) the DKGMCC algorithm when $o = 0$; (3) the DKMCC algorithm when $o = 0$ and $\xi = 2$; (4) the DKLMP algorithm when $\lambda \to 0_+, o = 0$; the DKLMS algorithm when $\lambda \to 0_+, o = 0$, and $\xi = 2$.

When the model has new input data $x_n(m)$ inputs, the kernel adaptive filter output can be expressed as:

$$d_n(m+1) = \boldsymbol{\omega}_n^{\text{T}}(m+1)\varphi(\boldsymbol{x}_n(m+1)) = \mu\sum_{i=1}^{m} g(e_n(i))\kappa(\boldsymbol{x}_n(i), \boldsymbol{x}_n(m+1)) \tag{9.13}$$

The value of the parameter o in the DKGMCC-VC algorithm is mainly used to capture the mean nature of the noise. In the presence of nonzero-mean noise, the value of the free parameter significantly affects the convergence and steady-state performance of the algorithm. In order to determine the optimal value of the parameter o, the algorithm needs to solve an objective function. This objective function is designed to optimize the performance of the algorithm in a nonzero-mean noise environment and also ensure that the algorithm's convergence and steady-state performance are optimal. By solving this objective function, the value of parameter o that best suits the current noise characteristics can be found, thus improving the overall performance of the DKGMCC-VC algorithm. To obtain the value of the most appropriate parameter o, the following objective function is solved:

$$o_n(m) = \arg\min_{o\in C}\left\{\gamma_{\xi,\theta} - E\left[G_{\xi,\theta}(e_n(m) - o)\right]\right\}, \tag{9.14}$$

where C is the full set of values of o. Taking the error median as the positioning center can simplify the operation:

$$o_n(m) = \mathrm{median}\{e_n(m - N + 1), \ldots, e_n(m)\} \tag{9.15}$$

The mean-based variable center algorithm takes into account all the noise values, and when impulsive noise is present, it has a significant impact on the localization center, thus reducing the robustness of the algorithm. To mitigate this problem, the DKGMCC-VC algorithm employs the error median as the adaptive center. The median is a robust statistical measure that inherently suppresses the effects of outliers by focusing on the central tendency of the error distribution. This choice ensures that the kernel center remains consistent with the dominant error pattern, even if the noise exhibits heavy-tailed or asymmetric properties. Based on the above derivation, the pseudo-code of DKGMCC-VC is described in Algorithm 9.1, where $\boldsymbol{\alpha}_{n,i}$ denotes the component of the ith coefficient of the nth node at moment m at the time of iteration.

Algorithm 9.2: Pseudo-code for DQKGMCC-VC algorithm

Input: Kernel bandwidth σ; Step-size μ; Kernel parameter λ; Shape parameter ζ; Quantization size ρ

Output: Coefficient vector $\boldsymbol{\alpha}$

/* Initialization */

$m \leftarrow 1,\ n \leftarrow 1,\ p \leftarrow 1$

while $n \le N$ **do**

  $o_n(1) \leftarrow 0$

  $\mathrm{X}_n(1) \leftarrow x_n(1)$

  $\alpha_n(1) \leftarrow 0$

  $n \leftarrow n + 1$

  $\beta(n) \leftarrow 0$

end

$n \leftarrow 1$

/* Computation */

while $m \le M$ **do**

while $n \leq N$ **do**

$\hat{d}_n(m) \leftarrow \sum_{i=1}^{P} \alpha_{n,i}(m-1)\kappa_\sigma(\chi_{n,i}(m-1), \boldsymbol{x}_n(m))$

$e_n(m) \leftarrow d_n(m) - \hat{d}_n(m)$

$\Lambda_n(m) \leftarrow e_n(m) - o_n(m-1)$

$g(e_n(m)) \leftarrow \exp(-\lambda|\Lambda_n(m)|^{\xi})|\Lambda_n(m)|^{\xi-1} sign(\Lambda_n(m))$, $n \leftarrow n+1$

if $dis(\boldsymbol{x}_n(m), \chi_n(m-1)) < \rho$ **then**

$\beta(n) \leftarrow \beta(n) + 1$

$\chi_n(m) \leftarrow \chi_n(m-1)$

$i \leftarrow \arg\min_{1 \leq i \leq P} \|\boldsymbol{x}_n(m) - \chi_{n,i}(m-1)\|$

$\hat{\alpha}_{n,i}(m) \leftarrow \hat{\alpha}_{n,i}(m-1) + \mu g(e_n(m))$

else

$\chi_n(m) \leftarrow \{\chi_n(m-1), \boldsymbol{x}_n(m)\}$

$\hat{\alpha}_n(m) \leftarrow \{\hat{\alpha}_n(m-1), \mu g(e_n(m))\}$

end

if $m < M$ **then**

$o_n(m+1) \leftarrow 0$

else

$o_n(m+1) \leftarrow median\{e_n(m-N+1), \cdots, e_n(m)\}$

end

end

$\alpha_n(m) \leftarrow \sum_{l \in N_n} a_{l,n}\hat{\alpha}_n(m)$

$m \leftarrow m+1, n \leftarrow 1$

end

return $\boldsymbol{\alpha}$

As the DKGMCC-VC algorithm employs an iterative coefficient vector update based on training the sequence data set, the algorithm's complexity and memory demand both increase continuously. This can be regarded as a real-time growing radial basis function (RBF) network. Consequently, the algorithm is not well-suited to scenarios that demand high real-time performance. This issue can be effectively addressed through the sparsification method, which can inhibit the continuous growth of the network model. This section will introduce the DKGMCC-VC with online vector quantization (VQ) method algorithm, namely the DQKGMCC-VC algorithm.

The algorithm quantifies the feature vector $\varphi(\boldsymbol{x}_n(m))$. So the weight update equation is expressed as

$$\boldsymbol{\phi}_n(m+1) = \boldsymbol{\omega}_n(m) + \mu_n g(e_n(m))\mathbb{Q}[\varphi(\boldsymbol{x}_n(m))] \tag{9.16}$$

$$\boldsymbol{\omega}_n(m+1) = \sum_{l \in N_n} a_{l,n}\boldsymbol{\phi}_l(m+1) \tag{9.17}$$

where $\mathbb{Q}[\cdot]$ is the quantization operator in RKHS. However, because it is in a high-dimensional space, this quantification method cannot be directly used in the DKGMCC-VC algorithm. Moreover, in the Gaussian kernel, the original input space and the feature space distance are positively correlated. So the quantitative operations are generally performed in the original space:

$$\boldsymbol{\phi}_n(m+1) = \boldsymbol{\omega}_n(m) + \mu_n g(e_n(m))\kappa(\boldsymbol{Q}[\varphi(\boldsymbol{x}_n(m))], \cdot), \tag{9.18}$$

where $Q[\cdot]$ is the quantization operator of the original space.

The online VQ method generally selects the appropriate dictionary by using the Euclidean measure of the input space, i.e.,

$$dis(\boldsymbol{x}_n(i), \boldsymbol{x}_n(j)) = \|\boldsymbol{x}_n(i) - \boldsymbol{x}_n(j)\| \tag{9.19}$$

When the distance between the input data and the dictionary is smaller than the quantized size ρ, it remains unchanged, and unlike the traditional sparsification method, it locally updates the coefficient closest to the center using "redundant" data. Otherwise, the input data is appended to the dictionary and the corresponding coefficients are assigned.

Also, considering the weight combination part of the distributed network, the quantization method is partially improved by introducing a new counting array that records the number of times when the input data is below the quantization threshold with respect to the dictionary, which facilitates the subsequent combining process and solves the potential problems that may arise during the combining phase of the distributed network.

9.3 Theoretical Analysis

9.3.1 *Stability Analysis*

This section will use the energy conservation relation to analyze the stability of the system. Considering the unknown system as a nonlinear regression model, the model of the system on node n can be written as

$$d_n(m) = \boldsymbol{\omega}_o^{\mathrm{T}}\varphi(\boldsymbol{x}_n(m)) + v_n(m), \tag{9.20}$$

the output error of the algorithm is expressed as:

$$e_n(m) = d_n(m) - \tilde{\boldsymbol{\omega}}_n(m-1)^{\mathrm{T}}\varphi(\boldsymbol{x}_n(m)) \tag{9.21}$$

bring Eq. (9.20) into (9.21) yields:

$$e_n(m) = v_n(m) + \tilde{\boldsymbol{\omega}}_n(m-1)^{\mathrm{T}}\varphi(\boldsymbol{x}_n(m)) = n_n(m) + e_{a,n}(m) \tag{9.22}$$

where $e_{a,n}(m)$ is the prior error, $\tilde{\boldsymbol{\omega}}_n(m-1) = \boldsymbol{\omega}_o - \boldsymbol{\omega}_n(m-1)$ is the weight deviation. Iterated expression for the bias was obtained by subtracting from $\boldsymbol{\omega}_o$ the sides of Eq. (9.11)

$$\tilde{\boldsymbol{\phi}}_n(m) = \tilde{\boldsymbol{\omega}}_n(m-1) + \mu g(e_n(m))\varphi(\boldsymbol{x}_n(m)), \tag{9.23}$$

the posterior error is defined as $e_{p,n}(m) = \tilde{\boldsymbol{\omega}}_n(m)^{\mathrm{T}}\varphi(x_n(m))$, combining with Eqs. (9.21) and (9.23), we obtain

$$e_{p,n}(m) = e_{a,n}(m) - \mu g(e_n(m))k_\sigma(\boldsymbol{x}_n(m), \boldsymbol{x}_n(m)), \tag{9.24}$$

that is

$$e_{p,n}(m) = e_{a,n}(m) + \left(\tilde{\boldsymbol{\phi}}_n(m) - \tilde{\boldsymbol{\omega}}_n(m-1)\right)^{\mathrm{T}}\varphi(\boldsymbol{x}_n(m)), \tag{9.25}$$

by elimination $g(e_n(m))$, we can get

$$\tilde{\boldsymbol{\phi}}_n(m) = \tilde{\boldsymbol{\omega}}_n(m-1) + \frac{\left(e_{p,n}(m) - e_{a,n}(m)\right)\varphi(\boldsymbol{x}_n(m))}{\kappa_\sigma(\boldsymbol{x}_n(m), \boldsymbol{x}_n(m))} \tag{9.26}$$

Squaring and taking expectation operations on both sides of Eq. (9.25), the following energy conservation relationship is obtained

$$E\left[\|\tilde{\boldsymbol{\phi}}_n(m)\|_F^2\right] + E\left[\frac{e_{a,n}^2(m)}{\kappa_\sigma(\boldsymbol{x}_n(m), \boldsymbol{x}_n(m))}\right] = E\left[\|\tilde{\boldsymbol{\omega}}_n(m-1)\|_F^2\right] + E\left[\frac{e_{p,n}^2(m)}{\kappa_\sigma(\boldsymbol{x}_n(m), \boldsymbol{x}_n(m))}\right], \tag{9.27}$$

where $E\left[\|\tilde{\phi}_n(m)\|_F^2\right]$ is the weight error power (WEP) in the feature space $\mathbb{F}$.

Remark 9.2: A Gaussian kernel is used in this paper, So the equation $\kappa_\sigma(x_n(m), x_n(m)) = 1$ holds.

Apply Eq. (9.25) into (9.27), we obtain:

$$E\left[\|\tilde{\boldsymbol{\phi}}_n(m)\|_F^2\right] = E\left[\|\tilde{\boldsymbol{\omega}}_n(m-1)\|_F^2\right] + \mu^2 E\left[g^2(e_n(m))\right] - 2\mu E\left[e_{a,n}(m)g(e_n(m))\right] \tag{9.28}$$

Convergence analysis and stability analysis can be further performed by the above derived equation. In order to ensure DKGMCC-VC (9.13) algorithm convergence, Eq. (9.28) can obtain the step-size μ to meet the following condition

$$\mu \leq \frac{2E\left[e_{a,n}g(e_n(m))\right]}{E\left[g^2(e_n(m))\right]} = \frac{2E\left[e_{a,n}\exp\left(-\lambda|\Lambda_n(m)|^{\xi}\right)|\Lambda_n(m)|^{\xi-1}\text{sign}(\Lambda_n(m))\right]}{E\left[\exp\left(-2\lambda|\Lambda_n(m)|^{\xi}\right)|\Lambda_n(m)|^{2(\xi-1)}\right]} \tag{9.29}$$

where $\Lambda_n(m) = e_n(m) - o$. In order for the step-size μ_n to satisfy the convergence principle of the algorithm, there should exist:

$$E\left[e_{a,n}\exp\left(-\lambda|\Lambda_n(m)|^{\xi}\right)|\Lambda_n(m)|^{\xi-1}\text{sign}(\Lambda_n(m))\right] > 0 \tag{9.30}$$

Therefore, the algorithm in this paper is theoretically convergent when the step-size μ_n satisfies the above relationship, and the final step-size convergence condition needs to be further verified through experiments.

9.3.2 *Steady-State Mean-Squared Analysis*

In the following, we investigate the steady-state mean square performance of the proposed DKGMCC-VC algorithm, for tractable analysis, we adopt the following assumptions:

A1: The measurement noise $v_n(m)$ is an independent, identically distributed process with a mean value of o.
A2: Prior error $e^2_{a,n}(m)$ is independent of measurement noise $v_n(m)$, and $E[e_n(m)]$.

The following steady-state analysis of the system model is conducted to define the global excess mean square error of the steady state:

$$\vartheta^{\text{global}}_{\text{EMSE}} = \frac{1}{N}\sum_{n=1}^{N}\lim_{m\to\infty}E\left[e^2_{a,n}(m)\right] \tag{9.31}$$

the local is:

$$\vartheta^{\text{local}}_{\text{EMSE}} = \lim_{m\to\infty}E\left[e^2_{a,n}(m)\right] \tag{9.32}$$

We will perform the analysis with the local error. The following common assumptions are generally given for further derivation:
It is easy to know when the system is stable

$$\lim_{m\to\infty}\|\tilde{\boldsymbol{\omega}}_n(m+1)\|^2_F = \lim_{m\to\infty}\|\tilde{\boldsymbol{\omega}}_n(m)\|^2_F \tag{9.33}$$

combined Eqs. (9.33) with (9.28), we obtain:

$$2E\left[e_{a,n}(m)g(e_n(m))\right] = \mu E\left[g^2(e_n(m))\right] \tag{9.34}$$

It is obvious that the error distribution is independent of time m when the system is stabilized:

$$2E\big[e_{a,n}g(e_n)\big] = \mu E\big[g^2(e_n)\big] \tag{9.35}$$

By Taylor Expansion [37], we obtain

$$g(e_n) = g\big(v_n + e_{a,n}\big) = g(v_n) + g'(v_n)e_{a,n} + \frac{1}{2}g''(v_n)e_{a,n}^2 + o\big(e_{a,n}^2\big) \tag{9.36}$$

where $g'(v_n)$ and $g''(v_n)$ represent the first and second order derivatives, $o\big(e_{a,n}^2\big)$ are third and higher order derivatives, $g'(v_n)$ and $g''(v_n)$ are written as

$$g'(v_n) = \exp\big(-\lambda|v_n - o|^{\xi}\big)|v_n - o|^{\xi-2}\big(\xi - 1 - \lambda\xi|v_n - o|^{\xi}\big), \tag{9.37}$$

$$\begin{aligned} g''(v_n) &= \exp\big(-\lambda|v_n - o|^{\xi}\big)|v_n - o|^{\xi-3} \\ &\quad\times \big((\xi-1)(\xi-2) - 3\lambda\xi(\xi-1)|v_n - o|^{\xi} + \lambda^2\xi^2|v_n - o|^{2\xi}\big) \end{aligned} \tag{9.38}$$

By assumption A1 and A2, when $E\big[o\big(e_{a,n}^2\big)\big]$ small enough, $E\big[e_{a,n}g(e_n)\big]$ and $E\big[g^2(e_n)\big]$ can be obtained

$$E\big[e_{a,n}g(e_n)\big] = E\big[e_a g(v_n) + g'(v_n)e_{a,n}^2 + o\big(e_{a,n}^2\big)\big] \approx E\big[g'(v_n)\big]\vartheta_{\text{EMSE}}^{\text{local}} \tag{9.39}$$

$$E\big[g^2(e_n)\big] \approx E\big[g^2(v_n)\big] + E\Big[g(v_n)g''(v_n) + \big|g'(v_n)\big|^2\Big]\vartheta_{\text{EMSE}}^{\text{local}} \tag{9.40}$$

Substitute Eqs. (9.39) and (9.40) into Eq. (9.35), we obtain:

$$\vartheta_{\text{EMSE}}^{\text{local}} = \frac{\eta E\big[g^2(e_n)\big]}{2E\big[g'(v_n)\big] - \eta E\big[g(v_n)g''(v_n) + |g'(v_n)|^2\big]}. \tag{9.41}$$

Combining the obtained Eq. (9.41) with Eqs. (9.37) and (9.38), we can deduce that:

$$\vartheta_{\text{EMSE}}^{\text{local}} = \frac{\eta E\big[\exp(-2\lambda|v_n - o|^{\xi}|v_n - o|^{2\xi-2})\big]}{2E\big[\exp\big(-\lambda|v_n - o|^{\xi}\big)|v_n - o|^{\xi-2}\big(\xi - 1 - \lambda\xi|v_n - o|^{\xi}\big)\big] - \eta E[\chi(v_n)]} \tag{9.42}$$

where

$$\begin{aligned} \chi(v_n) &= \exp\big(-2\lambda|v_n - o|^{\xi}\big)|v_n - o|^{2\xi-4} \\ &\quad\times \big((\xi-1)(2\xi-3) - 5\lambda\xi(\xi-1)|v_n - o|^{\xi} + 2\lambda^2\xi^2|v_n - o|^{2\xi}\big) \end{aligned} \tag{9.43}$$

When the step size is small enough, Eq. (9.43) can be written as:

$$\vartheta_{\mathrm{EMSE}}^{\mathrm{local}} \approx \frac{\eta E\left[\exp\left(-2\lambda|\Lambda|^{\xi}|\Lambda|^{2\xi-2}\right)\right]}{2E\left[\exp\left(-\lambda|\Lambda|^{\xi}\right)|\Lambda|^{\xi-2}\left(\xi-1-\lambda\xi|\Lambda|^{\xi}\right)\right]}, \tag{9.44}$$

where $\Lambda = v_n - o$. Finally, the excess mean square error (EMSE) is obtained:

$$\vartheta_{\mathrm{EMSE}}^{\mathrm{global}} \approx \frac{1}{N}\sum_{n=1}^{N} \frac{\eta E\left[\exp\left(-2\lambda|\Lambda|^{\xi}|\Lambda|^{2\xi-2}\right)\right]}{2E\left[\exp\left(-\lambda|\Lambda|^{\xi}\right)|\Lambda|^{\xi-2}\left(\xi-1-\lambda\xi|\Lambda|^{\xi}\right)\right]} \tag{9.45}$$

Remark 9.3 The above steady-state performance analysis is derived under the condition that the higher-order error and step-size terms are sufficiently small. When they cannot be ignored, the simulation results will not match the theoretical values well.

9.4 Computer Simulation Examples

This section verifies the accuracy of the theoretical model and the superiority of the proposed algorithm. In addition, we compare the effects of different algorithms in different noise environments. As shown in Fig. 9.1, we used a diffusion connected network consisting of ten nodes. The input signal variance and background noise variance for each node are shown in Fig. 9.2. Furthermore, we set the linear combination coefficients employing the Metropolis rule [38]:

$$a_{l,n} = \begin{cases} 1/\max(x_l, x_n), \ l \in N_n, l \neq n \\ 1 - \sum\limits_{l \in N_n} a_{l,n}, \quad l = n \\ 0, \quad \text{otherwise} \end{cases} . \tag{(9.46)}$$

9.4.1 *Steady-State Behavior*

First, we test the correctness of the proposed theoretical model. The required signal is generated by the following nonlinear system [38]:

$d_n(m) = a_0 \sin(\boldsymbol{x}_n(m)) + a_1\boldsymbol{x}_n(m-1) + a_2\boldsymbol{x}_n^2(m-2) + v_n(m)(9.47)$,

In the steady-state behavior simulation, in order to guarantee an almost uniform convergence rate, we set the step size to 0.8 for the DKLMS, DKMCC and DKMCC-VC algorithm, and 1.2 for the DKGMCC and DKGMCC-VC algorithms. Meanwhile, we choose the shape parameter ξ is 4, the kernel parameter λ is 2, and the kernel bandwidth σ is 1. All parameters are set by scanning for optimal results. Furthermore,

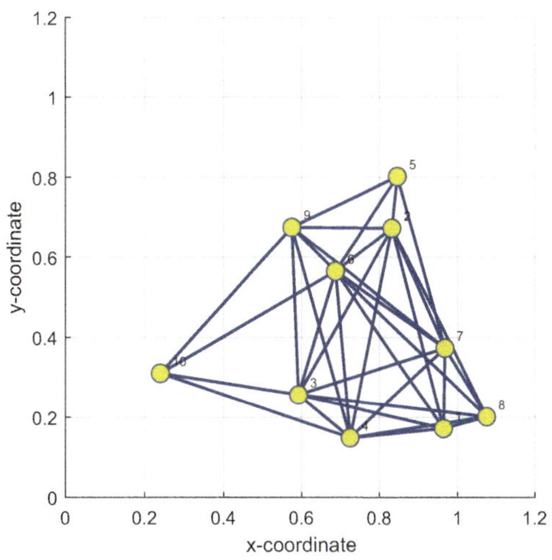

Fig. 9.1 Network topology of nodes

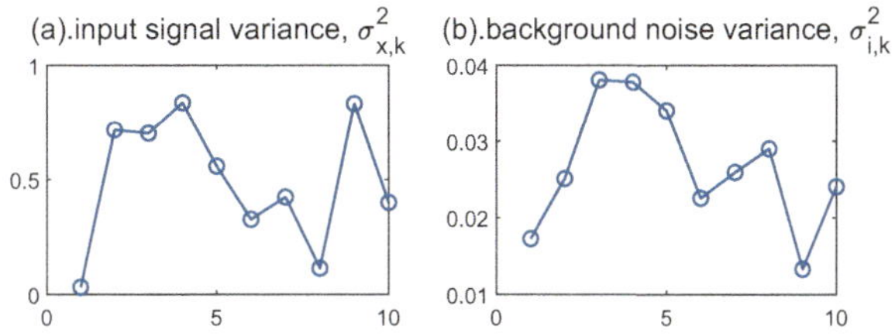

Fig. 9.2 Input signal variance and background noise variance plots

we choose a uniform distribution on [0, b] as the noise. And to ensure the stability and accuracy of the algorithm, a total of 100,000 iterations are performed and the average of the last 1000 iterations is taken as the simulated EMSE.

Figures 9.3, 9.4 and 9.5, respectively, show the relationship between the simulated EMSE and the theoretical EMSE for different step sizes, noise variances and center values, where b is the true mean parameter in the actual noise environment, independent of the algorithm settings, and o is defined as an adjustable parameter that is used to simulate the noise mean value set by the algorithm. From the figures, it can be seen that when the step size and variance are small, and when the center position o is close to the noise mean, the EMSE values between the theoretical model and the simulation model are basically the same. Furthermore, when the step size and variance gradually increase and the noise center is far away from the center position, the theoretical model cannot predict the behavior of the DKGMCC-VC algorithm well, and the steady-state EMSE becomes larger as the step size and variance increase and the noise center is far away.

Fig. 9.3 Simulated and theoretical EMSEs of DKGMC-VC algorithm with different step size μ ($b = 0.25$, $o = 0.25$)

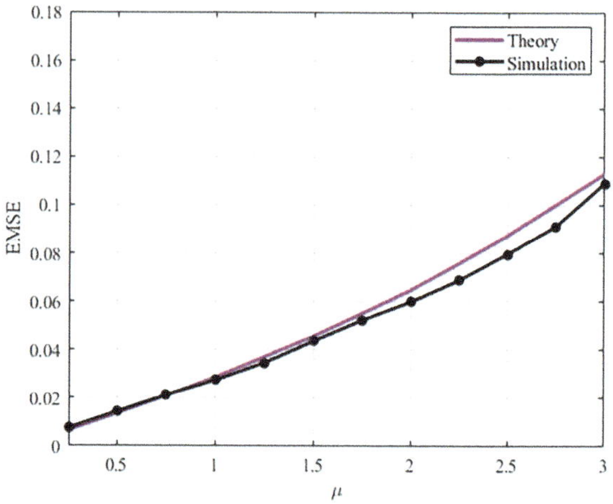

Fig. 9.4 Simulated and theoretical EMSEs of DKGMC-VC algorithm with different noise variances ($\mu = 0.5$, $o = 0.5$)

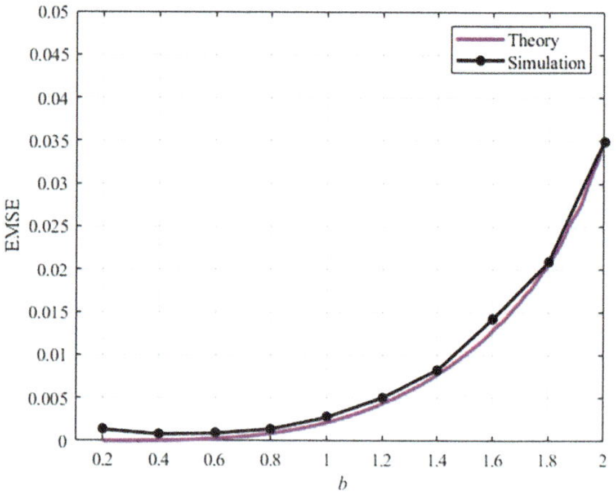

Fig. 9.5 Simulated and theoretical EMSEs of DKGMC-VC algorithm with different center location o ($\mu = 0.5$, $b = 1$)

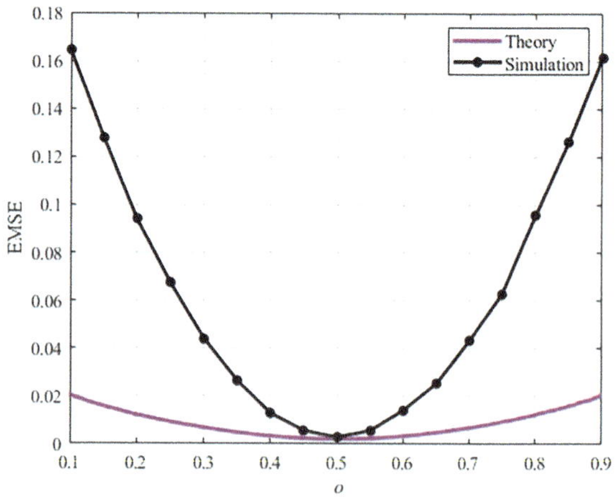

9.4.2 Performance Comparison

We use the global steady-state EMSE as the judgment index of algorithm performance, and obtained the EMSE learning curve of each node through 100 independent operations. The global EMSE is defined as follows:

$$\mathrm{EMSE}(m) = \frac{1}{N}\sum_{n=1}^{N} 10\log_{10}\left(e_{a,n}^{2}(m)\right). \tag{9.48}$$

Using this global EMSE expression, the proposed DKGMCC-VC algorithm is further compared and analyzed with other algorithms under distributed networks. During the simulation, the design kernel size σ is 1, the kernel parameter λ is 2. In addition, the noise mean is used to represent the center position. For different algorithms, choosing their most appropriate step size can lead to the convergence curve with the best performance. We tested the following four noises separately:

(1) Gaussian with mean of 1 and variance of 0.25.
(2) Uniform distribution on [0,2].
(3) Laplacian distribution with mean of 0.25 and variance of 0.25.
(4) Binary distribution over $\{0,1\}$ of the probability mass $p[v_n = 0] = p[v_n = 1] = 0.5$.

As illustrated in Fig. 9.6a–d, the convergence curves of EMSE obtained from 300 independent Monte Carlo experiments demonstrate that the DKGMCC-VC algorithm has superior performance compared to other algorithms in all four distinct noise environments. When the noise mean is nonzero, the performance of DKLMS, DKMCC and DKGMCC algorithms all degrade significantly, The DKGMCC-VC algorithm, because of its variable center property, can reduce the error through adjusting the center position. Meanwhile, the convergence speed of DKGMCC algorithm is not significantly affected, and it always maintains a fast convergence speed. These phenomena further demonstrate the superiority and robustness of the variable center algorithm.

9.4.3 Performance of the DQKGMCC-VC

At the third point of this section, the convergence performance of the DQKGMCC-VC algorithm is compared with the other algorithms. In the simulations, the central position is o and the kernel parameters are 1,0.5,2. Select noise with binary distribution over $\{0,1\}$, probability mass $p[v_n = 0] = p[v_n = 1] = 0.5$. Through relevant simulation results, the quantification size ρ is fixed to 0.5. The EMSE learning curve is shown in Fig. 9.7. The relationship between the network size and the number of iterations of the DKGMCC-VC and DQKGMCC-VC algorithms is shown in Fig. 9.8.

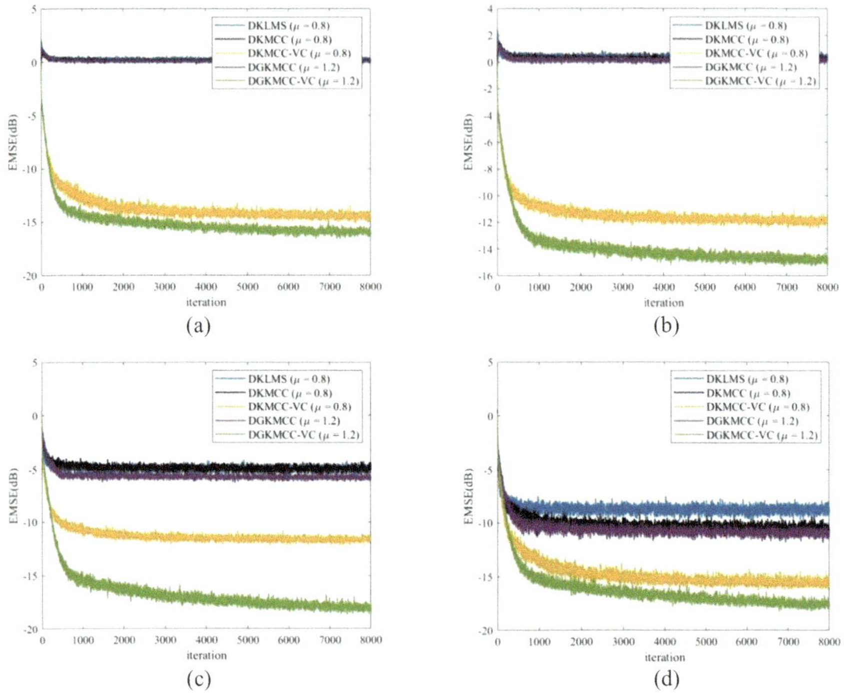

Fig. 9.6 Learning curves for the EMSE with different noise distributions: **a** Gaussian; **b** Uniform; **c** Binary; **d** Laplace

Figures 9.7 and 9.8 show that the EMSE learning curves of DQKGMCC-VC algorithm is almost the same as that of DKGMCC-VC algorithm, while the computational complexity is much lower than that of DKGMCC-VC algorithm. it is worth noting that when the number of iterations is further increased, the complexity advantage of the quantized algorithm is especially obvious, which fully demonstrates that the sparsification method can effectively reduce the computational complexity while maintaining the accuracy of the algorithm.

9.4.4 *Chaotic Time Series Prediction*

The Mackey–Glass (MG) chaotic time series is a classical nonlinear dynamical model that uses differential equations to generate discrete time series. The expression for this model used in [39] is:

$$\frac{du_{k,n}}{d_n} = \xi \frac{u_{k,n-\tau}}{1 + u_{k,n-\tau}^{a}} - \rho u_{k,n} \tag{9.49}$$

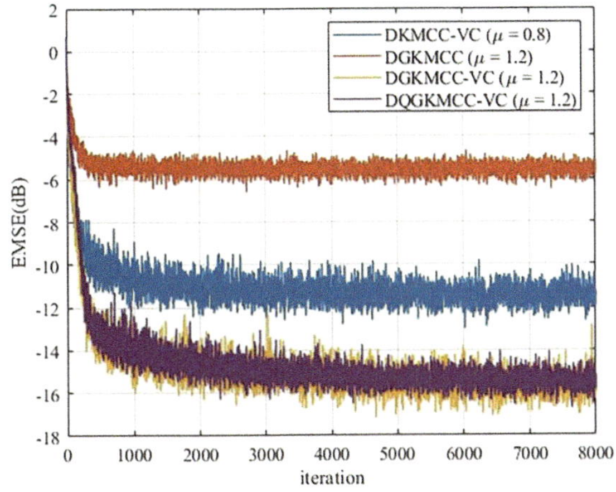

Fig. 9.7 EMSE convergence curves of DQKGMCC-VC and other algorithms

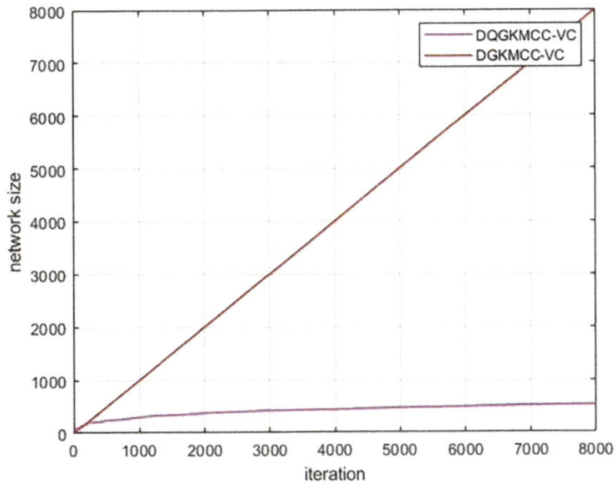

Fig. 9.8 Network size growth of DKGMCC-VC and KGMCC-VC algorithms

The dynamical model is governed by four critical parameters: a time delay constant $\tau = 0.3$, a delay effect coefficient $\xi = 0.2$, a nonlinearity scaling factor $a = 10$, and a damping coefficient $\rho = 0.1$. In a diffusion network, all nodes collaboratively predict output by employing last five input samples. The initial value of the sequence generated by each node is randomly selected from [1.1, 1.3]. Meanwhile, in order to fully realize the advantages of the proposed algorithm, we first generated noise-free MG time series and constructed the input–output dataset, and subsequently added nonzero-mean noise to the output data. And for performance evaluation, we use 2000 samples corrupted by noise as training data and 200 noise-free samples as testing data. The lowest steady-state EMSE was obtained for DKGMCC-VC, as illustrated in Fig. 9.9, which verifies the merit of DKGMCC-VC for nonlinear prediction under impulsive interference environments.

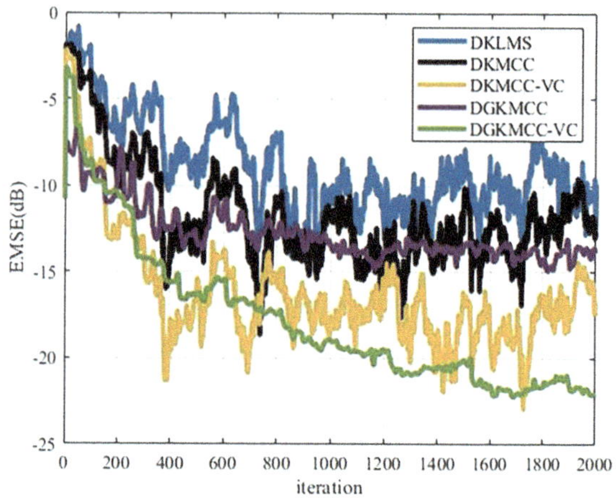

Fig. 9.9 MSE plots of chaotic time series prediction with different algorithms

9.5 Summary

In this paper, in order to improve the performance of distributed network estimation under nonlinear system model with nonzero-mean noise environment, a diffusion kernel adaptive filtering algorithm based on the GMCC-VC, i.e., the DKGMCC-VC algorithm, is proposed. The algorithm has lower steady-state performance compared to existing distributed network algorithms in nonzero noise environment. Meanwhile, the quantized DKGMCC-VC (DQKGMCC-VC) algorithm is proposed, which significantly reduces the complexity of the algorithm, improves the computational efficiency, and saves the computational space while ensuring the accuracy of the algorithm. Finally, the convergence conditions and steady-state performance of the DKGMCC-VC algorithm are analyzed based on the energy conservation relation. The final simulation results also further prove the correctness of the theoretical results. However, it can also be seen from the results that the theoretical and simulated values do not fit well when the higher-order error is large or the step size is long, which is a direction that needs attention in the future.

References

1. S.S. Haykin, *Adaptive Filter Theory* (Pearson Education India, 2008)
2. B. Widrow, E. Walach, Adaptive signal processing for adaptive control. IFAC Proc. Vol. **16**, 7–12 (1983)
3. W. Liu, P. P. Pokharel, J. C. Principe, The kernel least-mean-square algorithm. IEEE Trans. Signal Process. **56**(2), 544–554 (2008)
4. T.K. Paul, T. Ogunfunmi, Study of the convergence behavior of the complex kernel least mean square algorithm. IEEE Trans. Neural Netw. Learn. Syst. **24**(9), 1349–1363 (2013)
5. T.K. Paul, T. Ogunfunmi, A kernel adaptive algorithm for quaternion-valued inputs. IEEE Trans. Neural Netw. Learn. Syst. **26**(10), 2422–2439 (2015)

6. Y. Engel, S. Mannor, R. Meir, The kernel recursive least-squares algorithm. IEEE Trans. Signal Process. **52**(8), 2276–2285 (2004)
7. S. Van Vaerenbergh, M. Lazaro-Gredilla, I. Santamaria, Kernel recursive least-squares tracker for time-varying regression. IEEE Trans. Neural Netw. Learn. Syst. **23**(8), 1314–1326 (2012)
8. W. Liu, J.C. Príncipe, Kernel affine projection algorithms. EURASIP J. Adv. Signal Process. 1–12 (2008)
9. W. Liu, I. Park, Y. Wang, J.C. Principe, Extended kernel recursive least squares algorithm. IEEE Trans. Signal Process. **57**(10), 3801–3814 (2009)
10. Y. Zhu, H. Zhao, X. Zeng, B. Chen, Robust generalized maximum correntropy criterion algorithms for active noise control. IEEE/ACM Trans. Audio Speech Lang. Process. **28**, 1282–1292 (2020)
11. X. Hou, H. Zhao, X. Long, An innovative transient analysis of adaptive filter with maximum correntropy criterion. IEEE Signal Process. Lett. **29**, 1689–1693 (2022)
12. S. Zhao, B. Chen, J. C. Príncipe, Kernel adaptive filtering with maximum correntropy criterion, *The 2011 International Joint Conference on Neural Networks*, San Jose, CA, USA (2011), pp. 2012–2017
13. B. Chen, Z. Yuan, N. Zheng, J.C. Príncipe, Kernel minimum error entropy algorithm. Neurocomputing **121**, 160–169 (2013)
14. B. Shen, X. Wang, L. Zou, Maximum correntropy kalman filtering for non-gaussian systems with state saturations and stochastic nonlinearities. IEEE/CAA J. Automatica Sinica **10**(5), 1224–1233 (2023)
15. Y. Zhu, H. Zhao, X. He, Z. Shu, B. Chen, Cascaded random fourier filter for robust nonlinear active noise control. IEEE/ACM Trans. Audio Speech Lang. Process. **30**, 2188–2200 (2022)
16. A. H. Sayed, *Adaptation, Learning, and Optimization Over Networks*, vol. 7 (NOW, Boston, MA, USA, 2014)
17. A. H. Sayed, S. -Y. Tu, J. Chen, X. Zhao, Z. J. Towfic, Diffusion strategies for adaptation and learning over networks: an examination of distributed strategies and network behavior. IEEE Signal Process. Mag. **30**(3), 156–171 (2013)
18. A. Ainomäe, M. Bengtsson, T. Trump, Distributed largest eigenvalue-based spectrum sensing using diffusion LMS. IEEE Trans. Signal Inf. Process. Netw. **4**(2), 362–377 (2018)
19. S. Kanna, D. H. Dini, Y. Xia, S. Y. Hui, D. P. Mandic, Distributed widely linear Kalman filtering for frequency estimation in power networks. IEEE Trans. Signal Inf. Process. Netw. **1**(1), 46–57 (2015)
20. Y. Zhou, H. Zhao, D. Liu, X. Guo, Distributed active noise control robust to impulsive interference. IEEE Sensors J. (2025)
21. Y. Zhou, H. Zhao, D. Liu, M-estimate based diffusion active noise control algorithm over distributed networks and its performance analysis. Signal Process. **225**, 0166–1684 (2024)
22. F. S. Cattivelli, C. G. Lopes, A. H. Sayed, A diffusion RLS scheme for distributed estimation over adaptive networks, in *IEEE Workshop on Signal Processing Advances in Wireless Communications, SPAWC, Helsinki, Finland* (2007), pp. 1–5
23. F. S. Cattivelli, C. G. Lopes, A. H. Sayed, Diffusion recursive least-squares for distributed estimation over adaptive networks. IEEE Trans. Signal Process. **56**(5), 1866–1877 (2008)
24. A. Naeimi Sadigh, H. Zayyani, A proportionate robust diffusion recursive least exponential hyperbolic cosine algorithm for distributed estimation. IEEE Trans. Circ. Syst. II Express Briefs **69**(4), 2381–2385 (2022)
25. C.G. Lopes, A.H. Sayed, Diffusion least-mean squares over adaptive networks: formulation and performance analysis. IEEE Trans. Signal Process. **56**(7), 3122–3136 (2008)
26. F. S. Cattivelli, A. H. Sayed, Diffusion LMS strategies for distributed estimation. IEEE Trans. Signal Process. **58**(3), 1036–1048 (2010)
27. X. Zhao, A.H. Sayed, Performance limits for distributed estimation over LMS adaptive networks. IEEE Trans. Signal Process. **60**(10), 5107–5124 (2012)
28. H. Zayyani, Robust minimum disturbance diffusion LMS for distributed estimation. IEEE Trans. Circ. Syst. II Express Briefs **68**(1), 521–525 (2021)

29. Y. Liu, C. Li, Z. Zhang, Diffusion sparse least-mean squares over networks. IEEE Trans. Signal Process. **60**(8), 4480–4485 (2012)
30. A.H. Sayed, Adaptive networks. Proc. IEEE **102**(4), 460–497 (2014)
31. W. Ma, B. Chen, J. Duan, H. Zhao, Diffusion maximum correntropy criterion algorithms for robust distributed estimation. Dig. Signal Process. **58**, 10–19 (2016)
32. B. Chen, L. Xing, H. Zhao, N. Zheng, J. C. Príncipe, Generalized correntropy for robust adaptive filtering. IEEE Trans. Signal Process. **64**(13), 3377–3387 (2016)
33. F. Chen, X. Li, S. Duan, Diffusion generalized maximum correntropy criterion algorithm for distributed estimation over multitask network. Dig. Signal Process. **81**, 17–25 (2018)
34. K. Hu, W. Xiong, Z.-Y. Wang, H.C. So, C.-S. Leung, Robust 4.D AOA localization based on maximum correntropy criterion with variable center. IEEE Trans. Signal Process. **72**, 5021–5035 (2024)
35. W. Liu, P. P. Pokharel, J. C. Principe, Correntropy: properties and applications in non-Gaussian signal processing. IEEE Trans. Signal Process. **55**(11), 5287–5298 (2007)
36. N. Takahashi, I. Yamada, A. H. Sayed, Diffusion least-mean squares with adaptive combiners: formulation and performance analysis. IEEE Trans. Signal Process. **58**(9), 4796–4810 (2010)
37. V. H. Nascimento, J. C. M. Bermudez, Probability of divergence for the least-mean fourth algorithm. IEEE Trans. Signal Process. **54**(4), 1377–1385 (2006)
38. B. Chen, S. Zhao, P. Zhu, J.C. Príncipe, Mean square convergence analysis for kernel least mean square algorithm. Signal Process. **92**(11), 2624–2632 (2012)
39. S. Wang, Y. Zheng, S. Duan, L. Wang, H. Tan, Quantized kernel maximum correntropy and its mean square convergence analysis. Dig. Signal Process. **63**, 164–176 (2017)

Chapter 10
Graph Diffusion Kernel Maximum Correntropy

In recent years, graph signal processing has attracted much attention due to its ability to model irregular and interactive data generated by wireless sensor networks (WSNs). However, there is no practical method to deal with the problem of modeling nonlinear systems in non-Gaussian noise environments. Given that the maximum correntropy criterion (MCC) exhibits robustness to impulsive and non-Gaussian noise, this paper introduces it into the graph kernel adaptive filtering algorithm and develops a graph diffusion kernel MCC (GDKMCC) algorithm. To suppress the problem of the infinite growth of the filter coefficient vector of the proposed algorithm, a pre-trained dictionary (PD) method is used in this paper. In addition, the mean square transient behavior and the convergence condition of the GDKMCC-PD algorithm are also provided under some assumptions. Finally, the simulation results verify the superiority of the proposed algorithm in modeling nonlinear systems under non-Gaussian noise environments and the correctness of the theoretical model.

10.1 Introduction

For several decades, wireless sensor networks (WSNs) have been commonly utilized in location estimation, environmental monitoring, frequency estimation, and other fields [1–3]. WSNs are usually placed in a distributed manner by a large number of sensor nodes, which means that a lot of recorded data is irregular and interactive structure [4]. However, it is difficult to handle this complex data structure using traditional signal processing tools. Graph theory provides an effective way to solve the problem of modeling these data and the complex structure between them. The so-called graph is a structure consisting of vertices and edges, which can be used to describe the relationship between sensor nodes. Typical topological graphs include the Erdos–Renyi graph, cyclic graph, random geometric graph, and so on. The topological graph is used to characterize irregular data structures, which can extend the

H. Zhao et al., *Information Theoretic Learning-Based Filter*,
https://doi.org/10.1007/978-3-032-29623-8_10

classical signal processing methods to the field of graph signal processing (GSP). Currently, GSP has been widely used in sensor networks [5], image processing [6], and speech enhancement [7].

Adaptive filtering algorithms, an essential part of traditional signal processing, have powerful self-learning and self-tracking capabilities. Although linear adaptive filtering has an excellent performance in areas such as modeling linear systems, it has limited capability in nonlinear system modeling applications. In order to improve the modeling ability of nonlinear systems, scholars have proposed a class of kernel adaptive filtering (KAF) algorithms by combining kernel methods with linear adaptive filtering. The so-called KAF algorithm transforms the input signal from the original space to the reproducing kernel Hilbert space (RKHS) by feature mapping and then constructing a nonlinear mapping relationship between the input and output using the traditional linear adaptive filtering algorithm. Due to the powerful nonlinear mapping capability of the kernel method, the limitations of the traditional adaptive filtering algorithm in dealing with nonlinear problems can be significantly improved. The classical KAF algorithms include kernel least mean square (KLMS) [8], kernel recursive least square (KRLS) [9], kernel affine projection algorithm (KAPA) [10] and extended kernel recursive least squares (Ex-KRLS) [11]. However, the above-mentioned KAF algorithms based on the minimum mean square error (MMSE) criterion suffer from deteriorating performance in non-Gaussian noise environments. To solve this problem, several KAF algorithms based on information learning theory [12, 13] such as kernel maximum correntropy (KMC) [14, 15] and kernel minimum error entropy (KMEE) [16] have been proposed.

Given that graph edges can be used to expose the interrelationships between node signals through underlying graph topology, its internal structure can be effectively explored. The adaptive filtering algorithm has a simple structure and low computational complexity. The relationship between the input and output of complex structured data can be effectively modeled if the adaptive filtering algorithm is combined with the graph signal processing technique. Based on this idea, scholars have derived linear graph adaptive filtering algorithms in recent years [17, 18]. However, when the underlying system model is highly nonlinear, the linear graph adaptive filtering algorithms will not work. To solve this problem, kernel and functional link methods are used to design the nonlinear graph adaptive filtering algorithms [19–21]. However, the current nonlinear graph adaptive filtering algorithms are based on the MMSE criterion and do not effectively handle the non-Gaussian noise case. Therefore, the purpose of this paper is to design a nonlinear graph adaptive filtering algorithm that can effectively handle nonlinear modeling problems in a non-Gaussian noise environment. According to this purpose, we introduce the maximum correntropy criterion (MCC) and derive a nonlinear graph adaptive filtering algorithm in this paper, namely, graph diffusion kernel maximum correntropy criterion (GDKMCC). In order to reduce the computational complexity of the proposed algorithm, a pre-trained dictionary (PD) method is used in this paper. The convergence condition and transient behavior of GDKMCC-PD algorithm are analyzed.

10.2 Graph Signal Processing

Consider an undirected graph $G = \{V, E\}$, where V and E denote the set of nodes $V = \{1, \ldots, N\}$ and the set of edges. If nodes k and l are directly connected, $(k, l) \in E$. The nodes in the graph locally interact by the graph-shift operator $\boldsymbol{S} \in \mathbb{R}^{N \times N}$, whose entry $s_{k,l}$ is nonzero only if $k = l$ or $(k, l) \in E$. The signal $\boldsymbol{x}(m)$ on the graph at time m is defined as $\boldsymbol{x}(m) = [x_1(m), \ldots, x_N(m)]^{\mathrm{T}}$, where $x_n(m)$ denotes the vertex value (signal strength) of node n. The local linear combination of samples of neighboring nodes is done by the graph-shift operation $\boldsymbol{Sx}(m)$. In general, it is assumed that the shift operator $\boldsymbol{S}$ is known, so the linear shift-invariant graph filter is defined as

$$\boldsymbol{H} = \sum_{i=0}^{L-1} h_i \boldsymbol{S}^i \tag{10.1}$$

where $\left[h_0, h_1, \ldots, h_{L-1}\right]^{\mathrm{T}}$ is the coefficient vector of linear graph filter and L denotes the order of graph filter. A common graph filter model [19, 22] produces a graph-filtered vector $\boldsymbol{d}(m)$ by

$$\boldsymbol{d}(m) = \sum_{i=0}^{L-1} h_i \boldsymbol{S}^i \boldsymbol{x}(m-i) + \boldsymbol{v}(m) \tag{10.2}$$

where $\boldsymbol{v}(m) = [v_1(m), \ldots, v_N(m)]^{\mathrm{T}}$ is an independent and identically distributed zero-mean noise. In the common linear graph diffusion adaptive filtering, a linear mapping relationship is established using the spread information samples $\boldsymbol{r}_n(m) = \left[x_n(m), [\boldsymbol{Sx}(m-1)]_n, \ldots, \left[\boldsymbol{S}^{L-1}\boldsymbol{x}(m-L-1)\right]_n\right]^{\mathrm{T}}$ of node n and the model output $d_n(m)$. Unfortunately, it is difficult to fully capture the nonlinear relationship between inputs and outputs in most practical applications through a linear model. Therefore, we use a nonlinear model, namely

$$d_n(m) = f(\boldsymbol{r}_n(m)) + v_n(m) \tag{10.3}$$

To obtain the nonlinear mapping relation $f(\cdot)$, we use the kernel adaptive filtering algorithm in this paper.

10.3 Graph Diffusion Kernel Maximum Correntropy

In graph signal processing, N spread information samples are available at time m. Therefore, given the data pairs $\{\boldsymbol{r}_n(i), d_n(i)\}_{n=1,i=1}^{N,m-1} \bigcup \{\boldsymbol{r}_n(m), d_n(m)\}_{n=1}^{N}$, according to the representer theorem, the estimation of $y_n(m)$ in distributed implementation can be expressed as

$$\hat{d}_n(m) = f_{n,m}(\boldsymbol{r}_n(m)) = \sum_{i=1}^{m-1} \boldsymbol{\alpha}_{n,i}(m-1)k_\delta(\boldsymbol{r}_n(i), \boldsymbol{r}_n(m)) = \boldsymbol{\Omega}_n^{\mathrm{T}}(m)\boldsymbol{\psi}(\boldsymbol{r}_n(m)) \quad (10.4)$$

where $f_{n,m}(\cdot)$ is local estimate of $f(\cdot)$ at node n, $\boldsymbol{\Omega}_n(m) = f_{n,m}(\cdot) = \sum_{i=1}^{m-1} \boldsymbol{\alpha}_{n,i}(m-1)k_\delta(\boldsymbol{r}_n(i), \cdot)$, $\boldsymbol{\alpha}_{n,i}(m-1)$ is ith expansion coefficient, $\boldsymbol{\psi}(\cdot) : \mathbb{R}^L \rightarrow \mathbb{F}$ is the feature mapping from the original data space to RKHS $\mathbb{F}$, and $k_\delta(\cdot, \cdot)$ is kernel function in kernel method controlled by kernel bandwidth δ. Since the existing graph kernel adaptive filtering algorithms utilize the MMSE criterion as the cost function [19, 20], resulting in poor performance of the algorithms in the impulsive noise or non-Gaussian noise environment. To improve the performance of the algorithm, the MCC criterion is used in this paper [23, 24]. Therefore, the global cost function can be defined as

$$J_{\mathrm{MCC,glob}} = \sum_{n=1}^{N} J_n(m) \quad (10.5)$$

where $J_n(m) = \frac{1}{M}\sum_{i=1}^{M} k_\sigma\left(d_n(i) - \boldsymbol{\Omega}^{\mathrm{T}}(m)\boldsymbol{\psi}(\boldsymbol{r}_n(i))\right)$.

Without loss of generality, the kernel function in the correntropy uses Gaussian kernel function in this paper, that is

$$k_\sigma(e_n(i)) = \exp\left(-\frac{e_n^2(i)}{2\sigma^2}\right) \quad (10.6)$$

In the diffusion strategy, sensor nodes are able to exchange information with their neighboring nodes. Therefore, we decompose the global cost function into the local cost function

$$J_{\mathrm{MCC},n} = \sum_{l\in\mathcal{N}_n} a_{l,n} J_{l,n}(m) \quad (10.7)$$

where

$$J_{l,n}(m) = \frac{1}{M}\sum_{i=1}^{M} k_\sigma\left(d_l(i) - \boldsymbol{\Omega}_n^{\mathrm{T}}(m)\boldsymbol{\psi}(\boldsymbol{r}_l(i))\right) \quad (10.8)$$

and $\mathcal{N}_n$ denotes the set of neighboring nodes of node n, i.e., all nodes of the set connected to n by an edge and $a_{l,n}$ is combination coefficient that satisfies $\sum_{l\in\mathcal{N}_n} a_{l,n} = 1$ and $a_{l,n} \geq 0$. To obtain the optimal estimation $\boldsymbol{\Omega}_n(m)$, the stochastic information gradient method is used in this paper due to the low computational complexity of the adaptive filtering algorithm based on this method. At moment m, the instantaneous gradient of $J_{\mathrm{MCC},n}$ with respect to $\boldsymbol{\Omega}_n(m)$ is

$$\frac{\partial J_{\mathrm{MCC},n}}{\partial \boldsymbol{\Omega}_n(m)} = \frac{1}{\sigma^2} \sum_{l \in \mathcal{N}_n} a_{l,n} k_\sigma(e_l(m)) e_l(m) \boldsymbol{\psi}(\boldsymbol{r}_l(m)) \tag{10.9}$$

where $e_l(m) = d_l(m) - \boldsymbol{\Omega}_l^{\mathrm{T}}(m)\boldsymbol{\psi}(\boldsymbol{r}_l(m))$ because both $\boldsymbol{\Omega}_l^{\mathrm{T}}(m)$ and $\boldsymbol{\Omega}_n^{\mathrm{T}}(m)$ are the estimates of $f(\cdot)$ [25]. Thus, the update equation at node n is denoted as

$$\begin{aligned}
\boldsymbol{\Omega}_n(0) &= 0, \\
\boldsymbol{\Omega}_n(m+1) &= \boldsymbol{\Omega}_n(m) + \mu \sum_{l \in \mathcal{N}_n} a_{l,n} k_\sigma(e_l(m)) e_l(m) \boldsymbol{\psi}(\boldsymbol{r}_l(m)) \\
&= \boldsymbol{\Omega}_n(m-1) + \mu \sum_{i=m-1}^{m} \sum_{l \in \mathcal{N}_n} a_{l,n} k_\sigma(e_l(i)) e_l(i) \boldsymbol{\psi}(\boldsymbol{r}_l(i)) \\
&\cdots \\
&= \mu \sum_{i=1}^{m} \sum_{l \in \mathcal{N}_n} a_{l,n} k_\sigma(e_l(i)) e_l(i) \boldsymbol{\psi}(\boldsymbol{r}_l(i))
\end{aligned} \tag{10.10}$$

where $\mu = \eta/\sigma^2$ and η is the step size. At moment m, the optimal estimate at node n can be expressed as

$$\boldsymbol{\Omega}_n(m) = \sum_{l \in \mathcal{N}_n} a_{l,n} \widehat{\boldsymbol{\Omega}}_l(m) \tag{10.11}$$

where $\widehat{\boldsymbol{\Omega}}_l(m)$ is the intermediate estimate. Substituting (10.11) into (10.10) yields

$$\widehat{\boldsymbol{\Omega}}_l(m) = \mu \sum_{i=1}^{m} k_\sigma(e_l(i)) e_l(i) \boldsymbol{\psi}(\boldsymbol{r}_l(i)) \tag{10.12}$$

By comparing Eqs. (10.4) and (10.12), the update process of the expansion coefficients is written as

$$\widehat{\boldsymbol{\alpha}}_{l,m}(m) = \mu k_\sigma(e_l(m)) e_l(m) \tag{10.13}$$

and $\widehat{\boldsymbol{\alpha}}_{l,i}(m) = \widehat{\boldsymbol{\alpha}}_{l,i}(m-1)$ for all $i < m$.

According to the adapt-then-combine (ATC) method in the diffusion strategy [25], one can obtain an estimate $\boldsymbol{\alpha}_{n,m}(m)$ by

$$\begin{cases}
\widehat{\boldsymbol{\alpha}}_{n,m}(m) = \mu k_\sigma(e_n(m)) e_n(m) \\
\boldsymbol{\alpha}_{n,m}(m) = \sum_{l \in \mathcal{N}_n} a_{l,n} \widehat{\boldsymbol{\alpha}}_{l,m}(m)
\end{cases} \tag{10.14}$$

Based on the above derivation, we summarize the pseudo code of ATC-GDKMCC algorithm in Algorithm 10.1.

Algorithm 10.1: Pseudo Code for ATC-GDKMCC algorithm

```
Input: Kernel bandwidths σ, δ; Step size μ
Output: Coefficient vector α
/* Initialization                                              */
m ← 1, n ← 1
/* Computation                                                 */
while m ≤ M do
    while n ≤ N do
        if m = 1 then
            α_{n,1}(1) ← 0
        end
        d̂_n(m) ← Σ_{i=1}^{m-1} α_{n,i}(m-1) k_δ(r_n(i), r_n(m))
        e_n(m) ← d_n(m) - d̂_n(m)
        α_{n,m}(m) ← μ k_σ(e_n(m)) e_n(m)
        n ← n + 1
    end
    α_{n,m}(m) ← Σ_{l∈N_n} a_{l,n} α_{n,m}(m)
    m ← m + 1, n ← 1
end
return α
```

Given the fact that the computational complexity of the KAF algorithm increases with the number of input signals, to reduce the computational complexity, this paper considers a pre-trained dictionary (PD) method [26, 27]. Given that the dictionary size of node n is D, the kernelized input vector can therefore be written as

$$\boldsymbol{k}_n(m) = \left[k_\delta\left(\boldsymbol{r}_n(m), \boldsymbol{r}_{n,1}\right), \ldots, k_\delta\left(\boldsymbol{r}_n(m), \boldsymbol{r}_{n,D}\right)\right]^{\mathrm{T}} \tag{10.15}$$

where $\boldsymbol{r}_{n,d}$ denotes the dth dictionary component of node n and it is generated from the same probability density function as the spread information samples because common sparsification methods use inputs that satisfy certain criteria as dictionary components, such as approximate linear dependence [9], coherence-check criterion [19], and online vector quantization [14], etc. Of course, it can also be obtained by these sparsification methods. Based on a pre-trained dictionary, the output of the filter can be written as

$$\hat{d}_n(m+1) = \boldsymbol{\alpha}_n^{\mathrm{T}}(m)\boldsymbol{k}_n(m+1) \tag{10.16}$$

where $\boldsymbol{\alpha}_n(m) = \left[\alpha_{n,1}(m), \ldots, \alpha_{n,D}(m)\right]^{\mathrm{T}}$. Based on Eq. (10.9) and combined with the ATC strategy, the weight update equation of the ATC-GDKMCC-PD is written as

$$\begin{cases} \widehat{\boldsymbol{\alpha}}_n(m+1) = \boldsymbol{\alpha}_n(m) + \mu k_\sigma(e_n(m))e_n(m)\boldsymbol{k}_n(m) \\ \boldsymbol{\alpha}_n(m+1) = \sum\limits_{l\in\mathcal{N}_n} a_{l,n}\widehat{\boldsymbol{\alpha}}_l(m+1) \end{cases} \tag{10.17}$$

From Eq. (10.17), it can be seen that the algorithm is a linear-in-the-parameters nonlinear filtering algorithm. The input of the filter is a kernelized input and the order of the filter is determined by the dictionary size. The pseudo-code of the ATC-GDKMCC-PD algorithm is summarized in Algorithm 10.2.

Algorithm 10.2: Pseudo Code for ATC-GDKMCC-PD algorithm

Input: Kernel bandwidths σ, δ; Step size μ
Output: Coefficient vector $\boldsymbol{\alpha}$
/* Initialization */
$m \leftarrow 1,\ n \leftarrow 1$
/* Computation */
while $m \le M$ **do**
  while $n \le N$ **do**
    if $m = 1$ **then**
      $\boldsymbol{\alpha}_{n,1}(1) \leftarrow \mathbf{0}$
    end
    $\hat{d}_n(m) \leftarrow \boldsymbol{\alpha}_n^T(m-1)\boldsymbol{k}_n(m)$
    $e_n(m) \leftarrow d_n(m) - \hat{d}_n(m)$
    $\boldsymbol{\alpha}_n(m) \leftarrow \boldsymbol{\alpha}_n(m-1) + \mu k_\sigma(e_n(m))e_n(m)$
    $n \leftarrow n+1$
  end
  $\boldsymbol{\alpha}_{n,m}(m) \leftarrow \sum_{l\in\mathcal{N}_n} a_{l,n}\boldsymbol{\alpha}_{n,m}(m)$
  $m \leftarrow m+1,\ n \leftarrow 1$
end
return $\boldsymbol{\alpha}$

10.4 Theoretical Analysis

To facilitate the subsequent analysis, we define the network filter coefficient vector as $\boldsymbol{\alpha}_g = \mathbf{1}_N \otimes \boldsymbol{\alpha}_{\text{opt}}$ and the network estimated filter coefficient vector as $\boldsymbol{\alpha}_g(m) = \left[\boldsymbol{\alpha}_1^{\mathrm{T}}(m), \boldsymbol{\alpha}_2^{\mathrm{T}}(m), \ldots, \boldsymbol{\alpha}_N^{\mathrm{T}}(m)\right]^{\mathrm{T}}$ and the kernelized input matrix $\boldsymbol{U}(m) = \text{blockdiag}\{\boldsymbol{k}_1(m), \boldsymbol{k}_2(m), \ldots, \boldsymbol{k}_N(m)\}$, where $\mathbf{1}_N$ is a column vector of length N with each element being 1, blockdiag$\{\cdot\}$ is the block-diagonal-stacking operator, and $\otimes$ denotes Kronecker product. Thus, the nonlinear system and the weight update equation at the network level are written as

$$\boldsymbol{y}(m) = \boldsymbol{U}^{\mathrm{T}}(m)\boldsymbol{\alpha}_g + \boldsymbol{v}(m) \tag{10.18}$$

$$\boldsymbol{\alpha}_g(m+1) = \boldsymbol{\mathcal{A}}\big(\boldsymbol{\alpha}_g(m) + \mu \boldsymbol{M}\boldsymbol{U}(m)\boldsymbol{e}(m)\big) \tag{10.19}$$

where $\boldsymbol{M} = \mathrm{diag}\{k_\sigma(e_1(m)) \otimes \mathbf{1}_D, \ldots, k_\sigma(e_N(m)) \otimes \mathbf{1}_D\}$, $\boldsymbol{e}(m) = [e_1(m), e_2(m), \ldots, e_N(m)]^{\mathrm{T}}$, $\boldsymbol{\mathcal{A}} = \boldsymbol{A}^T \otimes \boldsymbol{I}_D$, $\boldsymbol{I}_D$ is the $D \times D$ identity matrix, and $[\boldsymbol{A}]_{l,n} = a_{l,n}$. Since $\boldsymbol{A}$ is a left stochastic matrix, $\boldsymbol{\alpha}_g = \boldsymbol{\mathcal{A}}\boldsymbol{\alpha}_g$ holds. Defining the weight error vector as $\widetilde{\boldsymbol{\alpha}}_g(m) = \boldsymbol{\alpha}_g(m) - \boldsymbol{\alpha}_g$, we have

$$\tilde{\boldsymbol{\alpha}}_g(m+1) = \boldsymbol{\mathcal{A}}\big(\tilde{\boldsymbol{\alpha}}_g(m) + \mu \boldsymbol{M}\boldsymbol{U}(m)\boldsymbol{e}(m)\big) \tag{10.20}$$

$$\boldsymbol{e}(m) = \boldsymbol{d}(m) - \hat{\boldsymbol{d}}(m) = \boldsymbol{v}(m) - \boldsymbol{U}^T(m)\tilde{\boldsymbol{\alpha}}_g(m) \tag{10.21}$$

Substituting (10.21) into (10.20) yields

$$\tilde{\boldsymbol{\alpha}}_g(m+1) = \boldsymbol{\mathcal{A}}\big[\tilde{\boldsymbol{\alpha}}_g(m) - \mu \boldsymbol{M}\boldsymbol{U}(m)\boldsymbol{U}^{\mathrm{T}}(m)\tilde{\boldsymbol{\alpha}}_g(m) + \mu \boldsymbol{M}\boldsymbol{U}(m)\boldsymbol{v}(m)\big] \tag{10.22}$$

To obtain the convergence conditions of the proposed algorithm as well as the transient behavior, the following common assumptions are adopted.

A1 The weight error vector $\tilde{\boldsymbol{\alpha}}_g(m)$, the kernelized input $\boldsymbol{U}(m)$, and the measurement noise $\boldsymbol{v}(m)$ are independent of each other.

A2 $\boldsymbol{v}(m)$ is an independent, identically distributed (i.i.d.) noise signal with zero mean and covariance $\boldsymbol{R}_v = \mathrm{diag}\{\sigma_{v,n}^2\}_{n=1}^N$.

A3 The input vector $\boldsymbol{x}(m)$ is an i.i.d. zero-mean Gaussian signal with covariance $\boldsymbol{R}_{xx} = E\big[\boldsymbol{x}(m)\boldsymbol{x}^{\mathrm{T}}(m)\big]$.

10.4.1 Convergence Analysis

Taking expectations on both sides of Eq. (10.22) leads to

$$E\big[\tilde{\boldsymbol{\alpha}}_g(m+1)\big] = \boldsymbol{\mathcal{A}}(\mathbf{I}_{ND} - \mu E[\boldsymbol{M}]\boldsymbol{\mathcal{R}}_{kk})E\big[\tilde{\boldsymbol{\alpha}}_g(m)\big] \tag{10.23}$$

where $\boldsymbol{\mathcal{R}}_{kk}$ is autocorrelation matrix of $\boldsymbol{U}(m)$.

$$\boldsymbol{\mathcal{R}}_{\mathbf{kk}} = E\big[\boldsymbol{U}(m)\boldsymbol{U}^{\mathrm{T}}(m)\big] \approx \begin{bmatrix} E\big[\boldsymbol{k}_1(m)\boldsymbol{k}_1^{\mathrm{T}}(m)\big] & \cdots & E[\boldsymbol{k}_1(m)]E\big[\boldsymbol{k}_N^{\mathrm{T}}(m)\big] \\ \vdots & \ddots & \vdots \\ E[\boldsymbol{k}_N(m)]E\big[\boldsymbol{k}_1^{\mathrm{T}}(m)\big] & \cdots & E\big[\boldsymbol{k}_N(m)\boldsymbol{k}_N^{\mathrm{T}}(m)\big] \end{bmatrix}. \tag{10.24}$$

Since the Gaussian kernel function is used in this paper, the (i, j)th entry of the matrix $E\left[\boldsymbol{k}_n(m)\boldsymbol{k}_n(m)^{\mathrm{T}}\right]$ and the ith entry of the vector $E[\boldsymbol{k}_n(m)]$ are calculated as shown in (10.25) and (10.26),

$$\begin{aligned}
\left[E\left[\boldsymbol{k}_n(m)\boldsymbol{k}_n^{\mathrm{T}}(m)\right]\right]_{i,j} &= E\left[\exp\left(-\frac{1}{2\delta^2}\left(\left\|\boldsymbol{r}_n(m)-\boldsymbol{r}_{n,i}\right\|^2+\left\|\boldsymbol{r}_n(m)-\boldsymbol{r}_{n,j}\right\|^2\right)\right)\right] \\
&= \exp\left(-\frac{1}{2\delta^2}\left(\left\|\boldsymbol{r}_{n,i}\right\|^2+\left\|\boldsymbol{r}_{n,j}\right\|^2\right)\right) \\
&\quad E\left[\exp\left(-\frac{1}{\delta^2}\left(\left\|\boldsymbol{r}_n(m)\right\|^2-\left(\boldsymbol{r}_{n,i}+\boldsymbol{r}_{n,j}\right)^{\mathrm{T}}\boldsymbol{r}_n(m)\right)\right)\right] \\
&\overset{a}{=}\left|\boldsymbol{I}_L+\frac{2}{\delta^2}\boldsymbol{R}_{rr,n}\right|^{-\frac{1}{2}} \\
&\quad \exp\left(-\frac{1}{4\delta^2}\left(2\left(\left\|\boldsymbol{r}_{n,i}\right\|^2+\left\|\boldsymbol{r}_{n,j}\right\|^2\right)-\left\|\boldsymbol{r}_{n,i}+\boldsymbol{r}_{n,j}\right\|^2_{\left(\boldsymbol{I}_L+\frac{\delta^2}{2}\boldsymbol{R}_{rr,n}^{-1}\right)^{-1}}\right)\right)
\end{aligned} \tag{10.25}$$

$$\begin{aligned}
[E[\boldsymbol{k}_n(m)]]_i &= E\left[\exp\left(-\frac{1}{2\delta^2}\left(\left\|\boldsymbol{r}_n(m)-\boldsymbol{r}_{n,i}\right\|^2\right)\right)\right] \\
&= \left|\boldsymbol{I}_L+\frac{1}{\delta^2}\boldsymbol{R}_{rr,n}\right|^{-\frac{1}{2}}\exp\left(-\frac{1}{8\delta^2}\left(4\left(\left\|\boldsymbol{r}_{n,i}\right\|^2\right)-\left\|\mathbf{r}_{n,i}\right\|^2_{\left(\boldsymbol{I}_L+\delta^2\boldsymbol{R}_{rr,n}^{-1}\right)^{-1}}\right)\right)
\end{aligned} \tag{10.26}$$

where $\boldsymbol{R}_{rr,n}=E\left[\boldsymbol{r}_n(m)\boldsymbol{r}_n(m)^{\mathrm{T}}\right]$.

The (m, n)th element of the covariance matrix of the spread information samples $\boldsymbol{r}_n(m)$ is obtained by

$$\begin{aligned}
\left[\boldsymbol{R}_{rr,n}\right]_{i,j} &= E\left[\left[\boldsymbol{S}^{i-1}\boldsymbol{x}(m-i+1)\right]_n\left[\boldsymbol{S}^{j-1}\boldsymbol{x}(m-j+1)\right]_n\right] \\
&= \operatorname{Tr}\left(\left[\boldsymbol{S}^{i-1}\right]_{n,\cdot}^{\mathrm{T}}\left[\boldsymbol{S}^{j-1}\right]_{n,\cdot}\boldsymbol{R}_{xx}\right)
\end{aligned} \tag{10.27}$$

To obtain the expectation of $\boldsymbol{M}$, we used the Taylor approximation, i.e., the mathematical expectation of its element $k_\sigma(e_n(m))$ is approximated as

$$E[k_\sigma(e_n(m))] \approx E\left[1-\frac{1}{2\sigma^2}v_n^2(m)\right]=1-\frac{\sigma_{v,n}^2}{2\sigma^2} \tag{10.28}$$

It is worth noting that when this approximation satisfies $\sigma \geq 1$, the higher order terms can be ignored. Therefore, if the algorithm converges, it needs to satisfy

$$\left\|\boldsymbol{\mathcal{A}}(\mathbf{I}_{\mathrm{ND}}-\mu E[\boldsymbol{M}]\boldsymbol{\mathcal{R}}_{kk})\right\| < 1 \tag{10.29}$$

Using the block maximum norm property leads to

$$\|\boldsymbol{\mathcal{A}}(\mathbf{I}_{ND} - \mu E[\boldsymbol{M}]\boldsymbol{\mathcal{R}}_{kk})\|_{b,\infty} \leq \|\mathcal{A}\|_{b,\infty}\|\boldsymbol{I}_{\text{ND}} - \mu E[\boldsymbol{M}]\boldsymbol{\mathcal{R}}_{kk}\|_{b,\infty} \tag{10.30}$$

Since $\boldsymbol{A}$ is left stochastic, $\|\boldsymbol{\mathcal{A}}\|_{b,\infty} = \left\|\boldsymbol{A}^{\mathrm{T}} \otimes \boldsymbol{I}_D\right\|_{b,\infty} = 1$ holds. Therefore, if the algorithm converges, it needs to satisfy

$$0 < \mu < \frac{2}{\lambda_{\max}(E[\boldsymbol{M}]\boldsymbol{\mathcal{R}}_{kk})} \tag{10.31}$$

10.4.2 Transient Performance Analysis

The transient performance analysis is accomplished by describing the evolution of the network mean square error (MSE), defined as

$$\text{MSE}(m) = \frac{1}{N}\sum_{n=1}^{N}\sigma_{v,n}^{2} + \text{EMSE}(m) \tag{10.32}$$

where $\text{EMSE}(m) = \frac{1}{N}E\left[\left\|\tilde{\boldsymbol{\alpha}}_g(m)\right\|_{\boldsymbol{\mathcal{R}}_{kk}}^2\right]$ is the network excess mean square error (EMSE). By

weighting, squaring and taking mathematical expectations on both sides of (10.22), we obtain

$$E\left[\left\|\tilde{\boldsymbol{\alpha}}_g(m+1)\right\|_{\mathcal{R}_{kk}}^2\right] = \mu^2 E\left[\left\|\boldsymbol{\mathcal{A}}\boldsymbol{M}\boldsymbol{p}_{uv}(m)\right\|_{\mathcal{R}_{kk}}^2\right] + E\left[\left\|\tilde{\boldsymbol{\alpha}}_g(m)\right\|_{\mathcal{B}^{\mathrm{T}}\mathcal{R}_{kk}\mathcal{B}}^2\right] \tag{10.33}$$

where $\mathcal{B} = \mathcal{A}(\mathbf{I}_{ND} - \mu E[\boldsymbol{M}]\mathcal{R}_{kk})$ and $\boldsymbol{p}_{uv}(m) = \boldsymbol{U}(m)\boldsymbol{v}(m)$. The first term on the right-hand side of (10.33) is written as [28]

$$E\left[\left\|\boldsymbol{\mathcal{A}}\boldsymbol{M}\boldsymbol{p}_{uv}(m)\right\|_{\mathcal{R}_{kk}}^2\right] = \text{Tr}(\mathcal{R}_{kk}\mathcal{G}) \tag{10.34}$$

where $\mathcal{G} = \boldsymbol{M}^{\mathrm{T}}\mathcal{A}^{\mathrm{T}}\mathcal{S}\mathcal{A}\boldsymbol{M}$ and $\mathcal{S} = E\left[\boldsymbol{p}_{uv}\boldsymbol{p}_{uv}^{\mathrm{T}}\right] = [\boldsymbol{R}_v \otimes \boldsymbol{I}_D]\mathcal{R}_{kk}$. Let $\boldsymbol{\Delta}_{kk} = \text{vec}(\mathcal{R}_{kk})$, (10.33) is rewritten as

$$E\left[\left\|\tilde{\boldsymbol{\alpha}}_g(m+1)\right\|_{\Delta_{kk}}^2\right] = E\left[\left\|\tilde{\boldsymbol{\alpha}}_g(m)\right\|_{\mathcal{F}\Delta_{kk}}^2\right] + \mu^2\text{vec}(\mathcal{G})^{\mathrm{T}}\boldsymbol{\Delta}_{kk} \tag{10.35}$$

where $\mathcal{F} \approx \mathcal{B}^{\mathrm{T}} \otimes \mathcal{B}^{\mathrm{T}}$ [28]. Iterating (10.35) from $m = 0$ yields

$$\text{EMSE}(m+1) = \frac{1}{N}E\left[\left\|\tilde{\boldsymbol{\alpha}}_g(0)\right\|_{\mathcal{F}^{m+1}\boldsymbol{\Delta}_{kk}}^2\right] + \frac{\mu^2}{N}\text{vec}(\mathcal{G})^{\mathrm{T}}\sum_{i=0}^{m}\mathcal{F}^i\boldsymbol{\Delta}_{kk} \tag{10.36}$$

Comparing EMSE($m+1$) with EMSE(m), the iterative relation of EMSE over time m can be obtained

$$\mathrm{EMSE}(m+1) = \mathrm{EMSE}(m) + \frac{\mu^2}{N}\mathrm{vec}(\mathcal{G})^{\mathrm{T}}\mathcal{F}^m\mathbf{\Delta}_{kk} + \frac{1}{N}E\left[\left\|\tilde{\boldsymbol{\alpha}}_g(0)\right\|^2_{(\mathcal{F}^{m+1}-\mathcal{F}^m)\mathbf{\Delta}_{kk}}\right] \tag{10.37}$$

According to Eq. (10.37), we can use it to predict the transient behavior of the ATC-GDKMCC-PD algorithm. If the statistical properties of the input signal and noise are known, we can directly obtain the convergence curve of the algorithm based on this model without training the algorithm.

10.5 Computer Simulation Examples

In this section, we use some simulated data and real data to verify the correctness of the theoretical analysis and the superiority of the proposed algorithm, respectively. Since the data between sensor nodes at short distances are correlated to some extent, we choose K sensors close to node k to build a topology to represent this correlation. if the nodes are connected, the entry $s_{k,l}$ of the shift operator $\boldsymbol{S}$ can be obtained by the following equation.

$$s_{k,l} = \exp\left(-\frac{d_{k,l}^2}{\overline{d}^2}\right) \tag{10.38}$$

where $d_{k,l}$ is Euclidean metric between nodes k and l, and $\overline{d}$ is the mean distance between all interconnected nodes.

10.5.1 Transient Performance

In this subsection, we evaluate the theoretical model of the ATC-GDKMCC-PD algorithm on a random connected graph topology with $N = 6$ sensor nodes and $K = 3$ shown in Fig. 10.1. The graph signal x_n is a zero-mean Gaussian process whose variance is generated from the uniform distribution in $[1, 1.5]$. The noise models v_n consider three different cases, namely, the uniform distribution over $[-b, b]$, the zero-mean Rayleigh distribution with variance $0.5(4-\pi)/4$, and the binary distribution over $\{-0.1, 0.9\}$ with probability $Pr(v_n(m) = 0.9) = 0.1$. And the combination coefficient $a_{l,n}$ are computed by $1/|\mathcal{N}_n|$, where $|\cdot|$ denotes the cardinality of its entry. The dictionary is a randomly generated $L \times D$ matrix, where $L = 2$ and $D = 11$ [27]. By default, the algorithm parameters in this subsection are set to $\sigma = 2$, $\delta = 0.8$, and $\eta = 0.8$. The learning curves are derived from averaging 50 independent experiments.

Fig. 10.1 Randomly connected graph topology with 6 sensor nodes

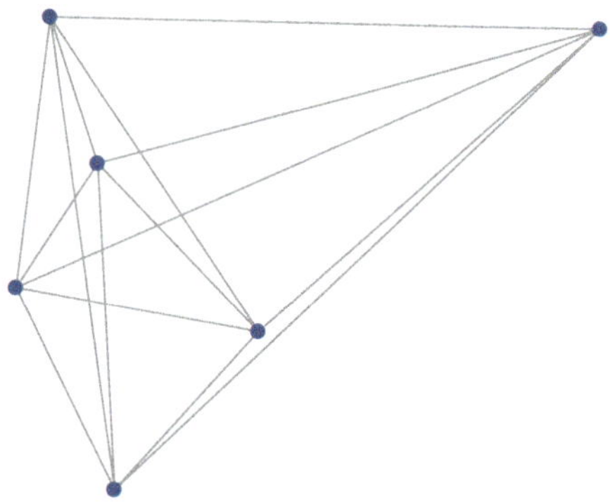

The theoretical and simulation results of the proposed algorithm with noise variance, step-size η, and kernel bandwidth σ for three different noise models are shown in Figs. 10.2, 10.3 and 10.4, where the dashed and solid lines show the simulation and the corresponding theoretical results generated by Eq. (10.37), respectively. It can be seen from Figs. 10.2, 10.3 and 10.4 that the proposed theoretical model is well agreed with the experimental results. As the variance of the noise decreases, the steady-state error of the algorithm is lower. In addition, the algorithm converges faster but its steady-state performance deteriorates as the step size increases and the kernel bandwidth decreases.

Fig. 10.2 Theoretical and simulation results of ATC-GDKMCC-PD algorithm over b in uniform noise environment

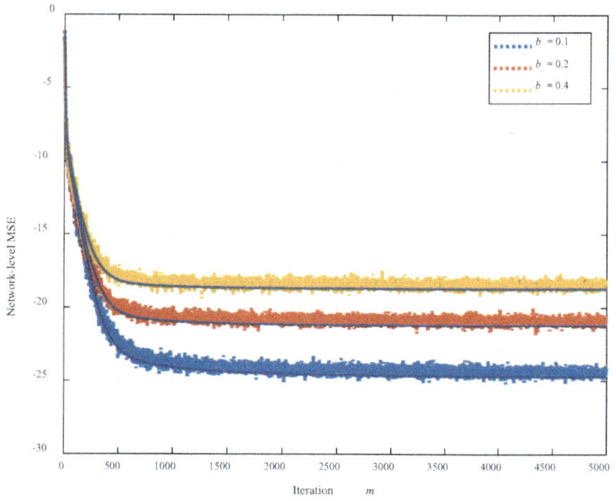

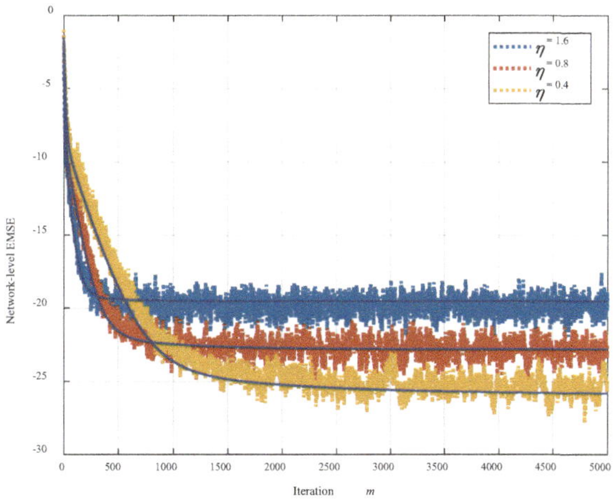

Fig. 10.3 Theoretical and simulation results of ATC-GDKMCC-PD algorithm over η in Rayleigh noise environment

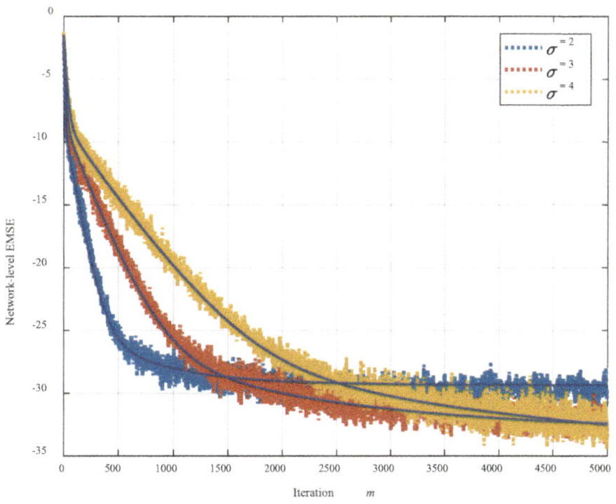

Fig. 10.4 Theoretical and simulation results of ATC-GDKMCC-PD algorithm over σ in binary noise environment

10.5.2 Nonlinear Graph Filter Identification

In this subsection, we test the performance of the ATC-GDKMCC and ATC-GDKMCC-PD algorithms using an example of modeling a nonlinear system under a non-Gaussian noise environment. The graph topology with $N = 20$ and $K = 3$ is given in Fig. 10.5. The noise $\mathbf{v}(m)$ is obtained from multiplicative model, defined as $v_n(m) = b_n(m)A_n(m)$, where $b_n(m)$ obeys Binary distribution over $\{0, 1\}$ with the probability mass function $Pr(b_n(m) = 1) = 0.5$, and $A_n(m)$ is an alpha-stable noise with the tail parameter $\alpha = 1.2$ and dispersion parameter $\gamma = 0.1$. The order of the filter is chosen to be 4, and the nonlinear model [19, 20] is given as

$$f(\boldsymbol{r}_n(m)) = \sqrt{r_{n,1}^2(m) + \sin^2\left(r_{n,4}(m)\pi\right)} + \left(0.8 - 0.5\exp\left(-r_{n,2}^2(m)\right)\right)r_{n,3}(m) \tag{10.39}$$

We compare the proposed algorithm with the linear adaptive graph filtering algorithms such as ATC strategy-based graph diffusion least mean square (ATC-GDLMS) and ATC strategy-based graph diffusion maximum correntropy (ATC-GDMCC) and nonlinear graph adaptive filtering algorithms such as ATC strategy-based graph diffusion kernel least mean square (ATC-GDKLMS) and ATC-GDKLMS with pre-trained dictionary (ATC-GDKLMS-PD) algorithms for network-level testing MSE as shown in Fig. 10.6. As we expected, the linear graph adaptive filtering algorithm fails in nonlinear modeling as shown in Fig. 10.6. The ATC-GDKMCC and ATC-GDKMCC-PD algorithms have better robustness in non-Gaussian noise environment compared to the ATC-GDKLMS and ATC-GDKLMS-PD algorithms. In addition, the steady-state performance of the full dictionary-based algorithms is slightly lower than that of the pre-trained dictionary-based algorithms.

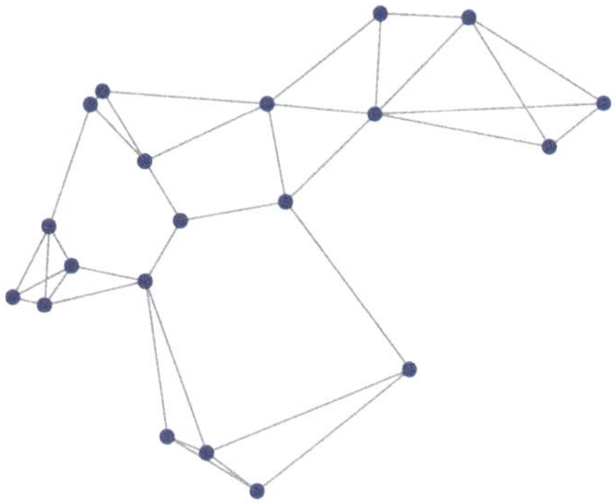

Fig. 10.5 Randomly connected graph topology with 20 sensor nodes

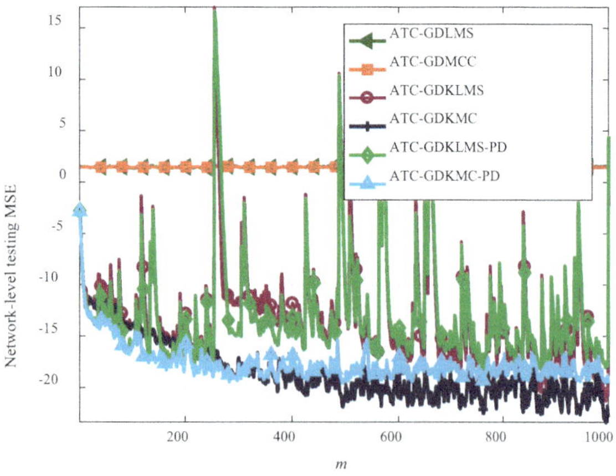

Fig. 10.6 Performance comparison of GDKMCC and the other algorithms in alpha-stable noise environment

10.5.3 Graph Signal Reconstruction for WSNs

In this subsection, we use an average dataset of July temperature values measured 299 temperature stations in Brazil from 1961 to 1990 as a static graph signal [29]. The graph topology with $K = 8$ of the original Brazilian temperature data is shown in Fig. 10.7. To test the reconstruction ability of the proposed algorithm for the graph signal, we add alpha-stable noise with $\alpha = 1.4$ and $\gamma = 0.1$ to the original graph signal. The parameters of the proposed algorithm are set as $\sigma = 2, \delta = 2, \mu = 0.01$, the filter order is chosen as $L = 4$. The reconstructed result by the ATC-GDKMCC algorithm after 1000 iterations are given in Fig. 10.8. It is easy to find that the proposed algorithm can effectively eliminate the contamination of the original data by noise and reconstruct the graph signal by comparing Figs. 10.7 and 10.8.

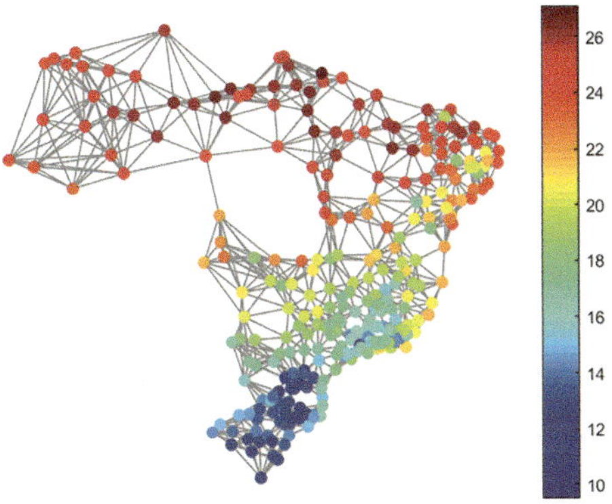

Fig. 10.7 Original graph signal from July data in Brazil temperature dataset

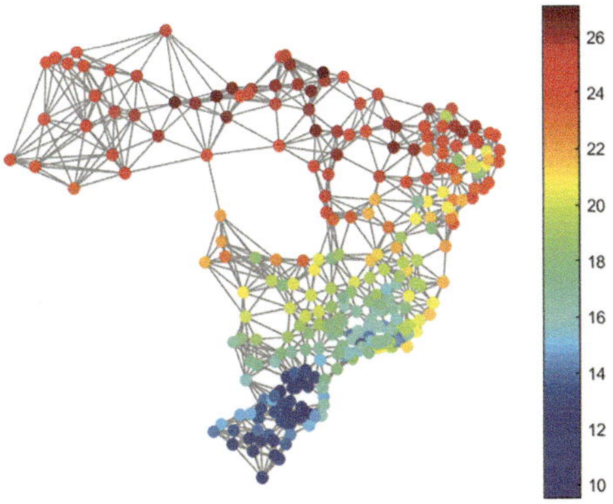

Fig. 10.8 Reconstructed graph signal by ATC-GDKMCC algorithm

10.5.4 Prediction in Temperature Sensor Networks

In this section, we evaluate the tracking performance of the proposed algorithm using the temperature data measured in the first 95 h of January 2010 at 205 U.S. temperature station. The graph topology with $K = 7$ of the 12th hour in the dataset is given in Fig. 10.9.

Figure 10.10 gives the temperature trends of the circled nodes in Fig. 10.9 over time and compares the tracking capabilities of the ATC-GDLMS ($\mu = 0.001$), ATC-GDMCC ($\mu = 0.001$, $\sigma = 2$), ATC-GDKLMS ($\mu = 0.3$, $\delta = 4$), and ATC-GDKMCC ($\mu = 0.3$, $\sigma = 2$, $\delta = 4$) algorithms. It can be seen from the Fig. 10.10 that the proposed algorithm is able to predict the sensor node temperature very well and the proposed algorithm has better accuracy compared with the other algorithms.

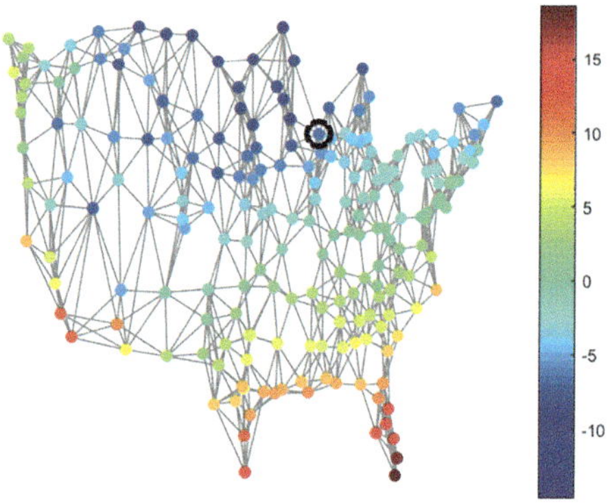

Fig. 10.9 Graph topology based on the 12th hour data in the U.S. dataset

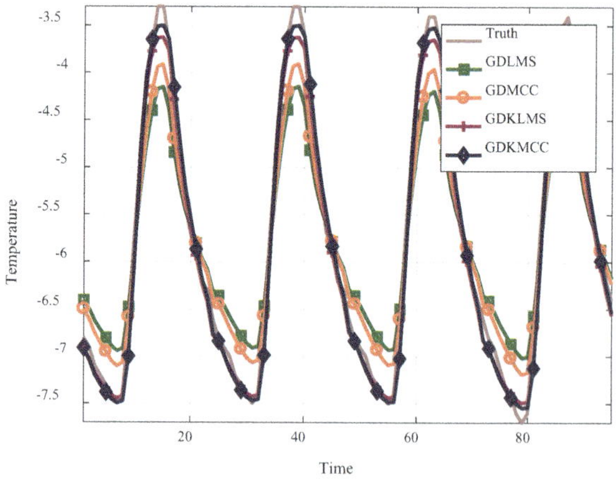

Fig. 10.10 Comparison of tracking capabilities of different algorithms

10.6 Summary

In this paper, we propose the graph diffusion KMCC algorithm based on the ATC strategy for dealing with the modeling problem of nonlinear systems of WSNs in a non-Gaussian noise environment. The effectiveness of the proposed algorithm is verified by nonlinear graph filter identification and prediction in temperature sensor networks. In addition, given that the KAF algorithm's computational complexity increases with the input data, we present ATC-GDMCC with a pre-trained dictionary algorithm to reduce the computational complexity of the proposed algorithm. Finally, we provide the convergence condition and transient performance analysis of the ATC-GDKMCC-PD algorithm, and the experimental results verify the correctness of the theoretical model.

References

1. H. Xiong, M. Peng, S. Gong, Z. Du, A novel hybrid RSS and TOA positioning algorithm for multi-objective cooperative wireless sensor networks. IEEE Sens. J. **18**(22), 9343–9351 (2018)
2. H.-C. Lin, Y.-C. Kan, Y.-M. Hong, The comprehensive gateway model for diverse environmental monitoring upon wireless sensor network. IEEE Sens. J. **11**(5), 1293–1303 (2011)
3. C. Li, H. Wang, Distributed frequency estimation over sensor network. IEEE Sens. J. **15**(7), 3973–3983 (2015)
4. W. Xu, H. Zhao, L. Zhou, Modified Huber M-estimate function-based distributed constrained adaptive filtering algorithm over sensor network. IEEE Sens. J. **22**(20), 19567–19582 (2022)
5. I. Jablonski, Graph signal processing in applications to sensor networks, smart grids, and smart cities. IEEE Sens. J. **17**(23), 7659–7666 (2017)
6. A.C. Yagan, M.T. Özgen, Spectral graph based vertex-frequency Wiener filtering for image and graph signal denoising. IEEE Trans. Signal Inf. Process. Netw. **6**, 226–240 (2020)
7. T. Wang, H. Guo, X. Yan, Z. Yang, Speech signal processing on graphs: the graph frequency analysis and an improved graph Wiener filtering method. Speech Commun. **127**, 82–91 (2021)
8. W. Liu, P.P. Pokharel, J.C. Principe, The kernel least-mean-square algorithm. IEEE Trans. Signal Process. **56**(2), 543–554 (2008)
9. Y. Engel, S. Mannor, R. Meir, The kernel recursive least-squares algorithm. IEEE Trans. Signal Process. **52**(8), 2275–2285 (2004)
10. W. Liu, J. C. Príncipe, Kernel affine projection algorithms. EURASIP J. Adv. Signal Process. **2008**(1), 1–12 (2008)
11. W. Liu, I. Park, Y. Wang, J.C. Principe, Extended kernel recursive least squares algorithm. IEEE Trans. Signal Process. **57**(10), 3801–3814 (2009)
12. Y. Zhu, H. Zhao, X. Zeng, B. Chen, Robust generalized maximum correntropy criterion algorithms for active noise control. IEEE/ACM Trans. Audio Speech Lang. Process. **28**, 1282–1292 (2020)
13. X. Hou, H. Zhao, X. Long, An innovative transient analysis of adaptive filter with maximum correntropy criterion. IEEE Signal Process. Lett. **29**, 1689–1693 (2022)
14. S. Wang, Y. Zheng, S. Duan, L. Wang, H. Tan, Quantized kernel maximum correntropy and its mean square convergence analysis. Digital Signal Processing **63**, 164–176 (2017)
15. X. Hou, H. Zhao, X. Long, W. Jin, The kernel recursive maximum total correntropy algorithm. IEEE Trans. Circ. Syst. II: Express Briefs **69**(12), 5139–5143 (2022)
16. B. Chen, Z. Yuan, N. Zheng, J.C. Príncipe, Kernel minimum error entropy algorithm. Neurocomputing **121**, 160–169 (2013)

17. P. Di Lorenzo, S. Barbarossa, P. Banelli, S. Sardellitti, Adaptive least mean squares estimation of graph signals. IEEE Trans. Signal Inf. Process. Netw. **2**(4), 555–568 (2016)
18. H. Zhao, W. Xiang, S. Lv, A variable parameter LMS algorithm based on generalized maximum correntropy criterion for graph signal processing. IEEE Trans. Signal Inf. Process. Netw. **9**, 140–151 (2023)
19. V.R.M. Elias, V.C. Gogineni, W.A. Martins, S. Werner, Adaptive graph filters in reproducing kernel Hilbert spaces: design and performance analysis. IEEE Trans. Signal Inf. Process. Netw. **7**, 62–74 (2021)
20. V. C. Gogineni, V. Naumova, S. Werner, Y. Huang, Graph kernel recursive least-squares algorithms, in *Asia-Pacific Signal and Information Processing Association Annual Summit and Conference (APSIPA ASC)* (2021), pp. 2072–2076
21. L. Li, Y.-F. Pu, Z.-Y. Luo, Distributed functional link adaptive filtering for nonlinear graph signal processing. Dig. Signal Process. **128**, 103558 (2022)
22. R. Nassif, C. Richard, J. Chen, A. H. Sayed, Distributed diffusion adaptation over graph signals, in *IEEE international conference on acoustics, speech, and signal processing (ICASSP)* (2018), pp. 4129–4133
23. W. Liu, P.P. Pokharel, J.C. Principe, Correntropy: Properties and applications in non-Gaussian signal processing. IEEE Trans. Signal Process. **55**(11), 5286–5298 (2007)
24. S. Zhao, B. Chen, J. C. Príncipe, Kernel adaptive filtering with maximum correntropy criterion, in *Proceedings of International Joint Conference Neural Networks* (2011), pp. 2012–2017
25. F.S. Cattivelli, A.H. Sayed, Diffusion LMS strategies for distributed estimation. IEEE Trans. Signal Process. **58**(3), 1035–1048 (2010)
26. W.D. Parreira, J.C.M. Bermudez, C. Richard, J. Tourneret, Stochastic behavior analysis of the Gaussian kernel least-mean-square algorithm. IEEE Trans. Signal Process. **60**(5), 2208–2222 (2012)
27. W. Gao, J. Chen, Transient performance analysis of zero-attracting Gaussian kernel LMS algorithm with pre-tuned dictionary. IEEE Access **7**, 135770–135779 (2019)
28. A. H. Sayed, Diffusion adaptation over networks, in *Academic Press Library in Signal Processing*, vol. 3 (O'Reilly, Boston, MA, USA, 2014), pp. 323–453
29. M. J. M. Spelta, W. A. Martins, Normalized LMS algorithm and data-selective strategies for adaptive graph signal estimation. Signal Process. **167**, 107326 (2020)

Chapter 11
Nonlinear Graph Diffusion Maximum Mixture Correntropy

In recent years, there has been a growing interest in graph signal processing due to its capability to model and analyze irregular data generated by wireless sensor networks (WSNs). Nevertheless, it is still an open problem to model graph nonlinear systems in non-Gaussian noise environments. Since common nonlinear adaptive filtering algorithms are developed based on the linear-in-the-parameters (LIP) model, this paper uses this model to design a family of nonlinear graph diffusion adaptive filtering algorithms to model graph nonlinear systems. In addition, taking into account the presence of non-Gaussian noise, this paper introduces the maximum mixture correntropy criterion (MMCC), which is generated by convex combination of two Gaussian kernel functions and exhibits more flexibility than the single kernel maximum correntropy criterion (MCC), and results in the nonlinear graph diffusion MMCC (NGD-MMCC) algorithm. The convergence conditions, mean-square transient behavior and mean-square steady-state behavior of the proposed algorithm are also obtained under some common assumptions. Finally, we verify the correctness of the theoretical analysis and the effectiveness of the proposed algorithm for modeling nonlinear systems in a non-Gaussian environment by simulated and measured data, respectively.

11.1 Introduction

In recent years, the large-scale deployment of wireless sensor networks (WSNs) has resulted in the generation of massive recorded data with irregular and interactive structures. In this context, graph signal processing methods have been rapidly developed. Currently, graph signal processing has been widely adopted in fields such as sensor networks, image processing, speech enhancement and social networks [1–4]. A graph is a mathematical structure consisting of vertices and edges used to describe pairwise relationships between vertices (sensor nodes), and graph signals are the data defined at the vertices of a graph [5]. Typical graph structures include

H. Zhao et al., *Information Theoretic Learning-Based Filter*,
https://doi.org/10.1007/978-3-032-29623-8_11

the Erdos–Renyi graph, the cyclic graph, the random geometric graph, and so on [6]. These structures provide the ability to handle these complex data structures, allowing traditional signal processing methods to be extended to the network domain.

In traditional signal processing, the least mean square (LMS) algorithm based on the minimum-mean-square-error (MMSE) criterion is one of the most popular adaptive filtering algorithms because it has low computational demands and is optimal under the Gaussian assumption [7]. However, while the LMS algorithm has excellent performance in areas such as linear system modeling, it has limited capability in modeling nonlinear systems. Several nonlinear adaptive filtering algorithms have been proposed to improve the algorithm's ability to model nonlinear systems, such as Volterra filters [8, 9], functional link adaptive filters [10–12], and kernel adaptive filters [13–15]. These nonlinear adaptive filtering algorithms perform various nonlinear expansions on the input signal and then use linear adaptive filtering algorithms to construct a nonlinear mapping relationship between the input and output. However, the above-mentioned algorithms perform poorly or even diverge in impulsive noise and non-Gaussian noise environments. Therefore, scholars have successively proposed various non-MMSE criteria [17, 18] including information theoretic learning (ITL) criterion to improve the performance of the algorithm [19]. Recently, the maximum correntropy criterion (MCC), an ITL-based criterion, has been successfully applied to robust adaptive filtering, which is known for its robustness in dealing with non-Gaussian noise [20, 21]. In order to improve the flexibility of the kernel function, the maximum mixture correntropy criterion (MMCC) is proposed, which utilizes a convex combination of two Gaussian kernels as the kernel function [22, 23].

The traditional adaptive filtering algorithms are limited in their ability to handle irregular and interactive data. Given that graphs can be used to explore the relationships between vertices through the graph structure, the internal structure of node signals can be effectively exploited.

The integration of the traditional adaptive filtering algorithms and graph signal processing methods enables the modeling of relationships between inputs and outputs of complex structured data. This combination inherits the characteristics of traditional adaptive filtering algorithms with low computational complexity and structural simplicity. In recent years, researchers have developed linear and nonlinear graph adaptive filtering algorithms to deal with the linear and nonlinear modeling of graph signals based on this concept [16, 24–28]. However, the classical nonlinear graph adaptive filtering algorithms are developed based on the MMSE criterion, resulting in their poor performance in non-Gaussian noise environments. To improve the performance of the algorithm in non-Gaussian noise environments, a robust algorithm is developed by introducing the maximum correntropy criterion into the graph kernel adaptive filtering algorithm [29]. Unfortunately, using the kernel method does not allow the algorithm to achieve optimal performance in arbitrary environments. In fact, the common nonlinear adaptive filtering algorithms satisfy the linear-in-the-parameters (LIP) model, in other words, they apply a nonlinear expansion to the input and then approximate the desired output with a linear combination of the expanded signals [30, 31]. Using the LIP model, the different nonlinear expansion methods can be utilized to approximate nonlinear systems and also provide a unified framework for nonlinear graph signal processing. Consequently, to improve the adaptability of the algorithm and the flexibility of the model, a class of nonlinear graph diffusion algorithms based on the LIP model is developed. To further improve the generality of the correntropy, the MMCC criterion is introduced and a class of nonlinear graph diffusion MMCC (NGD-MMCC) algorithm is developed in this paper. We also thoroughly analyze the convergence conditions of the proposed algorithm and comprehensively investigate its transient and steady-state statistical behaviors. In addition, we discuss the computational complexity of the different nonlinear expansions as well as the NGD-MMCC algorithm.

11.2 Linear-in-the-Parameters Model

It is clear from Eq. (10.2) that the filtered signal $y_n(m)$ at node n can be viewed as a linear combination of the spread graph signal $\mathbf{k}_n(m)$, i.e.,

$$d_n(m) = \mathbf{k}_n^{\mathrm{T}}(m)\mathbf{w}_o + v_n(m) \tag{11.1}$$

where $v_n(m)$ is the measurement noise, and

$$\begin{aligned}\mathbf{k}_n(m) &= \left[x_n(m), [\mathbf{S}\mathbf{x}(m-1)]_n, \ldots, \left[\mathbf{S}^{L-1}\mathbf{x}(m-L-1)\right]_n\right]^{\mathrm{T}} \\ &= \left[k_{n,1}(m), k_{n,2}(m), \ldots, k_{n,L}(m)\right]\end{aligned} \tag{11.2}$$

The linear relationship between the spread graph signal $\mathbf{k}_n(m)$ and the filtered output signal $d_n(m)$ can be effectively established by using different algorithms to obtain the coefficient vectors $\mathbf{w}_o$ in model (11.1). However, the linear model has limitations in dealing with complex nonlinear problems, and it is difficult to precisely capture the complex nonlinear relationship. To achieve this goal, we apply the LIP model [30, 31]. Thus, the nonlinear relationship between the input and output of node n based on this model can be written in

$$d_n(m) = \mathbf{q}_n^{\mathrm{T}}(m)\mathbf{h}_o + v_n(m) \tag{11.3}$$

where $\mathbf{h}_o$ is expansion coefficients and $\mathbf{q}_n(m)$ is a nonlinear expansion of the spread graph signal $\mathbf{k}_n(m)$. The expanded input vector $\mathbf{q}_n(m)$ can generally be expressed as follows:

$$\mathbf{q}_n(m) = \left[f_1^{\mathrm{T}}(\mathbf{k}_n(m)), f_2^{\mathrm{T}}(\mathbf{k}_n(m)), \ldots, f_M^{\mathrm{T}}(\mathbf{k}_n(m))\right]^{\mathrm{T}} \tag{11.4}$$

where M is the length of nonlinear expansion.

Remark 11.1 The class of filters represented by Eq. (11.3) is versatile and includes several commonly used nonlinear filtering algorithms. This model provides a flexible and powerful framework for nonlinear signal processing tasks in applications.

Based on the type of nonlinear expansion, these filtering algorithms can be classified as the functional link network (FLN) filtering, the Volterra filtering, and the random Fourier feature-based kernel adaptive filtering, etc. The extended input of FLN filtering based on trigonometric expansion can be written as [11]:

$$\mathbf{q}_n(m) = \big[\mathbf{k}_n^{\mathrm{T}}(m), \sin\big(\pi \mathbf{k}_n^{\mathrm{T}}(m)\big), \cos\big(\pi \mathbf{k}_n^{\mathrm{T}}(m)\big), \\ \ldots, \sin\big(B\pi \mathbf{k}_n^{T}(m)\big), \cos\big(B\pi \mathbf{k}_n^{\mathrm{T}}(m)\big)\big]^{\mathrm{T}} \tag{11.5}$$

where B is the expansion order. The extended input of FLN filtering based on exponential expansion is given as [12]:

$$\mathbf{q}_n(m) = \big[\mathbf{k}_n^{\mathrm{T}}(m), \exp\big(-a\big|\mathbf{k}_n^{\mathrm{T}}(m)\big|\big)\sin\big(\pi \mathbf{k}_n^{\mathrm{T}}(m)\big), \exp\big(-a\big|\mathbf{k}_n^{\mathrm{T}}(m)\big|\big)\cos\big(\pi \mathbf{k}_n^{\mathrm{T}}(m)\big), \\ \ldots, \exp\big(-a\big|\mathbf{k}_n^{\mathrm{T}}(m)\big|\big)\sin\big(B\pi \mathbf{k}_n^{\mathrm{T}}(m)\big), \exp\big(-a\big|\mathbf{k}_n^{\mathrm{T}}(m)\big|\big)\cos\big(B\pi \mathbf{k}_n^{\mathrm{T}}(m)\big)\big]^{\mathrm{T}} \tag{11.6}$$

where a is the exponential parameter and $|\cdot|$ is the absolute value operation. The extended input of the Volterra filtering is expressed as [8]:

$$\mathbf{q}_n(m) = \big[\mathbf{k}_n^{\mathrm{T}}(m), \mathbf{k}_n^{\mathrm{T}}(m) \odot \mathbf{k}_n^{\mathrm{T}}(m), \ldots, \mathbf{k}_n^{\mathrm{T}}(m) \odot \mathbf{k}_n^{\mathrm{T}}(m) \odot \cdots \odot \mathbf{k}_n^{\mathrm{T}}(m)\big]^{\mathrm{T}} \tag{11.7}$$

where $\odot$ is the Hadamard product. The random Fourier feature-based kernel adaptive filtering uses different Fourier feature functions to construct a random Fourier feature space. The extended input based on the cosine feature function is denoted as [32]:

$$\mathbf{q}_n(m) = \sqrt{\frac{2}{M}}\big[\cos\big(\boldsymbol{\omega}_1^{\mathrm{T}}\mathbf{k}_n(m) + b_1\big), \ldots, \cos\big(\boldsymbol{\omega}_M^{\mathrm{T}}\mathbf{k}_n(m) + b_M\big)\big]^{\mathrm{T}} \tag{11.8}$$

where $\boldsymbol{\omega}_i$ is drawn from the multivariate Gaussian distribution $\mathcal{N}\big(\mathbf{0}, \sigma^2\mathbf{I}_L\big)$ and b_i is sampled from the uniform distribution $\mathcal{U}(0, 2\pi)$. The extended input based on the Gaussian feature function is presented as [19, 33]:

$$\mathbf{q}_n(m) = \left[\exp\left(-\frac{\|\mathbf{k}_n(m) - \boldsymbol{\omega}_1\|^2}{2\sigma^2}\right), \ldots, \exp\left(-\frac{\|\mathbf{k}_n(m) - \boldsymbol{\omega}_M\|^2}{2\sigma^2}\right)\right]^{\mathrm{T}} \tag{11.9}$$

where $\boldsymbol{\omega}_i$ is generated from the input data using the same probability density function, σ is kernel bandwidth, and $||\cdot||^2$ is the squared Euclidean norm.

11.3 Nonlinear Graph Diffusion Maximum Mixture Correntropy Algorithm

The adaptive filtering algorithm based on the minimum mean square error (MMSE) criterion is ineffective due to the effect of non-Gaussian or impulsive noise in the system model. In order to improve the performance of the algorithm, the maximum mixture correntropy criterion (MMCC) is introduced in this paper. The mixture correntropy is a convex combination of two Gaussian kernels, which is more flexible compared to the correntropy of a single kernel. The cost function $J_n(\mathbf{h})$ of node n is defined as

$$J_n(\mathbf{h}) = E\left[\sum_{k=1}^{2} \lambda_k \exp\left(-\frac{\left(d_n(m) - \mathbf{q}_n^{\mathrm{T}}(m)\mathbf{h}\right)^2}{2\sigma_k^2}\right)\right] \tag{11.10}$$

where $E[\cdot]$ is the expectation operator, λ_1 and λ_2 are the mixture parameters and satisfy $0 \leq \lambda_1, \lambda_2 \leq 1$ and $\lambda_1 + \lambda_2 = 1$, σ_1 and σ_2 denote kernel bandwidths with $\sigma_1 \leq \sigma_2$. It is easy to observe from Eq. (11.10) that the mixture correntropy reduces to the traditional correntropy when λ_1 or λ_2 is equal to zero. To obtain the optimal expansion coefficient $\mathbf{h}$ in a sensor network, one can define the global cost function in the form of the sum of the cost functions of all nodes, i.e.,

$$J_{\text{glob}} = \sum_{l=1}^{N} J_l(\mathbf{h}) \tag{11.11}$$

The optimization problem can be rewritten using a diffusion strategy, where each node receives information from its neighbors and linearly fuses it with its own information, and the global cost function can be decomposed into a local cost function, that is,

$$J_{\text{loc},n} = \sum_{l \in \mathcal{N}_n} a_{l,n} J_l(\mathbf{h}) \tag{11.12}$$

Algorithm 11.1: Pseudo Code for NGD-MMCC algorithm

Input: Graph signal $\boldsymbol{x}(m)$, desired signal $y(m)$
Output: Expansion coefficient vector $\boldsymbol{h}$
/* Initialization */
σ_1, σ_2, λ_1, λ_2, η_n, $\boldsymbol{S}$, $a_{l,n}$, L
$m \leftarrow 1$, $n \leftarrow 1$
/* Computation */
while $m \leq M$ **do**
  while $n \leq N$ **do**
    if $m = 1$ **then**
      $\boldsymbol{h}_n(1) \leftarrow \boldsymbol{0}$
    end
    $\boldsymbol{k}_n(m) \leftarrow \left[x_n(m), \ldots, \left[\boldsymbol{S}^{L-1}\boldsymbol{x}(m-L-1)\right]_n\right]^T$
    $\boldsymbol{q}_n(m) \leftarrow \left[f_1^T(\boldsymbol{k}_n(m)), \ldots, f_M^T(\boldsymbol{k}_n(m))\right]^T$
    $\hat{d}_n(m) \leftarrow \boldsymbol{q}_n^T(m)\boldsymbol{h}_n(m)$
    $e_n(m) \leftarrow d_n(m) - \hat{d}_n(m)$
    $g(e_n(m)) \leftarrow \sum_{k=1}^{2} \frac{\lambda_k}{\sigma_k^2} \exp\left(-\frac{e_n^2(m)}{2\sigma_k^2}\right)$
    $\boldsymbol{\varphi}_n(m+1) \leftarrow \boldsymbol{h}_n(m) + \eta_n g(e_n(m)) e_n(m) \boldsymbol{q}_n(m)$
    $n \leftarrow n+1$
  end
  $\boldsymbol{\alpha}_{n,m}(m) \leftarrow \sum_{l \in \mathcal{N}_n} a_{l,n} \boldsymbol{\alpha}_{n,m}(m)$
  $m \leftarrow m+1$, $n \leftarrow 1$
end
return $\boldsymbol{h}$

where $\mathcal{N}_n$ is the set of neighboring nodes of node n and $a_{l,n}$ is combination coefficient between nodes l and n that satisfies $\sum_{l \in \mathcal{N}_n} a_{l,n} = 1$ and $a_{l,n} > 0$. A common method for optimizing the local cost function is to use the gradient descent method. In order to perform the gradient descent method, one requires to calculate the gradient of the local cost function with respect to $\mathbf{h}$, i.e.,

$$\begin{aligned}\frac{\partial J_{\mathrm{loc},n}}{\partial \mathbf{h}} &= \sum_{k=1}^{2} \sum_{l \in \mathcal{N}_n} \frac{\lambda_k a_{l,n}}{\sigma_k^2} E\left[\exp\left(-\frac{e_l^2(m)}{2\sigma_k^2}\right) e_l(m)\mathbf{q}_l(m)\right] \\ &\approx \sum_{k=1}^{2} \sum_{l \in \mathcal{N}_n} \frac{\lambda_k a_{l,n}}{\sigma_k^2} \exp\left(-\frac{e_l^2(m)}{2\sigma_k^2}\right) e_l(m)\mathbf{q}_l(m), \end{aligned} \tag{11.13}$$

where $e_l(m) = d_l(m) - \mathbf{q}_l^{\mathrm{T}}(m)\mathbf{h}_n(m)$. After obtaining the gradient $\frac{\partial J_{\mathrm{loc},n}}{\partial \mathbf{h}}$, we can use the gradient descent update rule to update the estimate of the expansion coefficient $\mathbf{h}$ at node n, say,

$$\mathbf{h}_n(m+1) = \mathbf{h}_n(m) + \eta_n \sum_{k=1}^{2} \sum_{l \in \mathcal{N}_n} \frac{\lambda_k a_{l,n}}{\sigma_k^2} \exp\left(-\frac{e_l^2(m)}{2\sigma_k^2}\right) e_l(m)\mathbf{q}_l(m) \tag{11.14}$$

where η_n is the positive step-size. According to the adapt-then-combine (ATC) method in the diffusion strategy [34], the current estimate $\mathbf{h}_n(m)$ and the intermediate estimate $\varphi_n(m)$ satisfy

$$\mathbf{h}_n(m) = \sum_{l \in \mathcal{N}_n} a_{l,n}\varphi_l(m) \tag{11.15}$$

Substituting Eq. (11.15) into Eq. (11.14) yields

$$\varphi_l(m+1) = \varphi_l(m) + \eta_n g(e_l(m)) e_l(m)\mathbf{q}_l(m) \tag{11.16}$$

where $g(e_l(m)) = \sum_{k=1}^{2} \lambda_k/\sigma_k^2 \exp\left(-e_l^2(m)/2\sigma_k^2\right)$. To facilitate the operation, we replace $\mathbf{h}_n(m)$ with $\mathbf{h}_l(m)$ in $e_l(m)$ in Eq. (11.16), that is, $e_l(m) = d_l(m) - \mathbf{q}_l^{\mathrm{T}}(m)\mathbf{h}_l(m)$. In addition, since $\mathbf{h}_n(m)$ contains more information, $\varphi_l(m)$ is replaced by $\mathbf{h}_l(m)$. Therefore, the weight update of the NGD-MMCC algorithm is given as

$$\begin{cases} \varphi_n(m+1) = \mathbf{h}_n(m) + \eta_n g(e_n(m)) e_n(m)\mathbf{q}_n(m) \\ \mathbf{h}_n(m+1) = \sum_{l \in \mathcal{N}_n} a_{l,n}\varphi_l(m+1) \end{cases} \tag{11.17}$$

Based on the above derivation, the NGD-MMCC algorithm is summarized in Algorithm 11.1.

11.4 Theoretical Analysis

In this section, we analyze the transient and steady-state behaviors of the NGD-MMCC algorithm using statistical theory. In addition, we also analyze the computational complexity of the proposed algorithm. First, we define the coefficient-error vector $\tilde{\mathbf{h}}_n(m)$ and the intermediate estimate coefficient-error $\tilde{\varphi}_n(m)$ of node n as:

$$\tilde{\mathbf{h}}_n(m) = \mathbf{h}_o - \mathbf{h}_n(m) \tag{11.18}$$

$$\tilde{\varphi}_n(m) = \mathbf{h}_o - \varphi_n(m) \tag{11.19}$$

Then the error $e_n(m)$ can be written as

$$e_n(m) = d_n(m) - \mathbf{q}_n^{\mathrm{T}}(m)\mathbf{h}_n(m) = \mathbf{q}_n^{\mathrm{T}}(m)\tilde{\mathbf{h}}_n(m) + v_n(m) \tag{11.20}$$

Without loss of generality, we define $\mathbf{q}_n(m) \in \mathbb{R}^P$. Substituting Eqs. (11.18), (11.19) and (11.20) into (11.17) yields

$$\begin{cases} \tilde{\varphi}_n(m+1) = \left[\mathbf{I}_P - \eta_n g(e_n(m))\mathcal{Q}_n(m)\right]\tilde{\mathbf{h}}_n(m) - \eta_n g(e_n(m))\mathbf{c}_n(m) \\ \tilde{\mathbf{h}}_n(m+1) = \sum\limits_{l\in\mathcal{N}_n} a_{l,n}\tilde{\varphi}_l(m+1), \end{cases} \tag{11.21}$$

where $\mathcal{Q}_n(m) = \mathbf{q}_n(m)\mathbf{q}_n^{\mathrm{T}}(m)$, $\mathbf{c}_n(m) = v_n(m)\mathbf{q}_n(m)$ and $\mathbf{I}_P$ is the $P \times P$ identity matrix. To facilitate our analysis, we will introduce some mathematical notation. Specifically, we use $\tilde{\mathbf{h}}_g(m) = \mathrm{col}\left\{\tilde{\mathbf{h}}_1(m), \tilde{\mathbf{h}}_2(m), \ldots, \tilde{\mathbf{h}}_N(m)\right\}$ to denote the network-level coefficient-error vector and $\tilde{\varphi}_g(m) = \mathrm{col}\{\tilde{\varphi}_1(m), \tilde{\varphi}_2(m), \ldots, \tilde{\varphi}_N(m)\}$ to denote the coefficient-error vector of the intermediate estimate at network level, where $\mathrm{col}\{\cdot\}$ denotes creation of a column vector by successively staking its arguments. Therefore, the global coefficient-error update equation can be rewritten as

$$\tilde{\mathbf{h}}_g(m+1) = \mathcal{B}(m)\tilde{\mathbf{h}}_g(m) - \mathcal{M}\mathcal{G}(m)\mathcal{A}\mathbf{c}(m) \tag{11.22}$$

where $\mathcal{M} = \mathrm{blockdiag}\{\eta_1\mathbf{I}_P, \eta_2\mathbf{I}_P, \ldots, \eta_N\mathbf{I}_P\}$, $\mathcal{A} = \mathbf{A}^{\mathrm{T}} \otimes \mathbf{I}_P$, $\mathbf{A}$ is a left stochastic matrix with $[\mathbf{A}]_{l,n} = a_{l,n}$, $\mathbf{c}(m) = \mathrm{col}\{\mathbf{c}_1(m), \mathbf{c}_2(m), \ldots, \mathbf{c}_N(m)\}$, $\mathcal{Q}(m) = \mathrm{blockdiag}\{\mathcal{Q}_1(m), \mathcal{Q}_2(m), \ldots, \mathcal{Q}_N(m)\}\mathcal{G}(m) = \mathrm{blockdiag}\{g(e_1(m))\mathbf{I}_P, g(e_2(m))\mathbf{I}_P, \ldots, g(e_N(m))\mathbf{I}_P\}$, $\mathcal{B}(m) = \mathcal{A}(\mathbf{I}_{NP} - \mathcal{M}\mathcal{G}(m)\mathcal{Q}(m))$, $\mathrm{blockdiag}\{\cdot\}$ is the block-diagonal-stacking operator, and $\otimes$ denotes Kronecker product.

In order to investigate the statistical behaviors and convergence conditions of the algorithm, we make the following assumptions:

A1 The coefficient-error vector $\tilde{\mathbf{h}}_g(m)$, the expanded input vector $\mathbf{q}_n(m)$, and the measurement noise $v_n(m)$ are independent of each other [25].
A2 The $v_n(m)$ at different nodes are mutually independent noise signals with zero mean and variance $\sigma_{v,n}^2$ [27].
A3 The graph signal $\mathbf{x}(m)$ is an independent, identically distributed zero-mean Gaussian signal with covariance $\mathbf{R}_{xx} = E\left[\mathbf{x}(m)\mathbf{x}^{\mathrm{T}}(m)\right]$ [35].
A4 The term $g(e_n(m))$ is independent of the other signals [36].

These assumptions are often used to analyze the performance of adaptive filtering algorithms. They allow us to analyze the algorithm under specific conditions, to obtain the behavior in different situations, and to explore the impact of different parameters on the performance of the algorithm.

11.4.1 Mean Convergence Analysis

According to assumption A3, we obtain that $\mathbf{k}_n(m)$ is also zero-mean Gaussian distribution and its covariance satisfies

$$\begin{aligned}\left[\mathbf{R}_{kk,n}\right]_{i,j} &= E\left[\left[\mathbf{S}^{i-1}\mathbf{x}(m-i+1)\right]_n\left[\mathbf{S}^{j-1}\mathbf{x}(m-j+1)\right]_n\right] \\ &= \mathrm{Tr}\left(\left[\mathbf{S}^{i-1}\right]_{n,\cdot}^{\mathrm{T}}\left[\mathbf{S}^{j-1}\right]_{n,\cdot}^{\mathrm{T}}\mathbf{R}_{xx}\right)\end{aligned} \tag{11.23}$$

where $\mathrm{Tr}(\cdot)$ is the trace of a matrix. We use the exponential expansion method as an example to explore the convergence conditions and statistical behavior of the algorithm. The performance analysis based on the nonlinear expansion of the Gaussian feature function can be found in [29]. The covariance matrix of the expanded input vector $\mathbf{q}_n(m)$ can be expressed as

$$\begin{aligned}\mathcal{R}_n = E[\mathcal{Q}_n(m)] \approx \mathrm{diag}\big\{&E\big[k_n^2(m)\big], E\big[k_n^2(m-1)\big], \ldots, E\big[k_n^2(m-L+1)\big], \\ &E\big[e^{-2a|k_n(m)|}\sin^2[\pi k_n(m)]\big], \ldots, E\big[e^{-2a|k_n(m-L+1)|}\sin^2[\pi k_n(m-L+1)]\big], \\ &E\big[e^{-2a|k_n(m)|}\cos^2[\pi k_n(m)]\big], \ldots, E\big[e^{-2a|k_n(m-L+1)|}\cos^2[\pi k_n(m-L+1)]\big], \\ &\ldots \\ &E\big[e^{-2a|k_n(m)|}\sin^2[B\pi k_n(m)]\big], \ldots, E\big[e^{-2a|k_n(m-L+1)|}\sin^2[B\pi k_n(m-L+1)]\big], \\ &E\big[e^{-2a|k_n(m)|}\cos^2[B\pi k_n(m)]\big], \ldots, E\big[e^{-2a|k_n(m-L+1)|}\cos^2[B\pi k_n(m-L+1)]\big]\big\}\end{aligned} \tag{11.24}$$

where $\mathrm{diag}\{\cdot\}$ is the diagonal operation.

In order to obtain the mathematical expectations in Eq. (11.24), the de Moivre theorem is introduced in this paper [37], i.e.,

$$\cos(Bx) + i\sin(Bx) = (\cos(x) + i\sin(x))^B = e^{iBx} \tag{11.25}$$

$$\cos(Bx) - i\sin(Bx) = (\cos(x) - i\sin(x))^B = e^{-iBx} \tag{11.26}$$

where $i = \sqrt{-1}$ denotes the imaginary unit. Based on Eqs. (11.25) and (11.26), we have

$$\sin^2(Bx) = \frac{2 - e^{2iBx} - e^{-2iBx}}{4} \tag{11.27}$$

$$\cos^2(Bx) = \frac{e^{2iBx} + e^{-2iBx} + 2}{4} \tag{11.28}$$

Assume that x obeys a zero-mean Gaussian distribution with variance σ^2. Therefore, the variance of the signal after nonlinear expansion can be written as

$$E\left[e^{-2ax}\sin^2(B\pi x)\right]=\frac{1}{2\sqrt{2\pi}\sigma}\left(2\int_0^\infty e^{-\frac{x^2}{2\sigma^2}-2ax}dx-\int_0^\infty e^{-\frac{x^2}{2\sigma^2}-2ax+2iB\pi x}dx\right.$$
$$\left.-\int_0^\infty e^{-\frac{x^2}{2\sigma^2}-2ax-2iB\pi x}dx\right) \tag{11.29}$$

$$E\left[e^{-2ax}\cos^2(B\pi x)\right]=\frac{1}{2\sqrt{2\pi}\sigma}\left(2\int_0^\infty e^{-\frac{x^2}{2\sigma^2}-2ax}dx+\int_0^\infty e^{-\frac{x^2}{2\sigma^2}-2ax+2iB\pi x}dx\right.$$
$$\left.+\int_0^\infty e^{-\frac{x^2}{2\sigma^2}-2ax-2iB\pi x}dx\right) \tag{11.30}$$

To solve the integral in Eqs. (11.29) and (11.30), we consider a general form of the integral of the following form:

$$\begin{aligned}\int \exp(-ax^2+bx+c)dx &= \int \exp\left[-a\left(x-\frac{b}{2a}\right)^2+\frac{b^2+4ac}{4a}\right]dx\\ &=\exp\left(\frac{b^2+4ac}{4a}\right)\int \exp\left[-\left(\sqrt{a}\left(x-\frac{b}{2a}\right)\right)^2\right]dx\\ &=\exp\left(\frac{b^2+4ac}{4a}\right)\frac{1}{\sqrt{a}}\int \exp(-t^2)dx,\ a>0\end{aligned} \tag{11.31}$$

where $t=\sqrt{a}\left(x-\frac{b}{2a}\right)$. To obtain a concrete solution for the integral, we now substitute the variables into the general form we derived earlier. Since it is difficult to get its exact value directly, we introduce the error function. The results of integration of the parts of Eqs. (11.29) and (11.30) are shown in Eqs. (11.32), (11.33) and (11.34).

$$\int e^{-\frac{x^2}{2\sigma^2}-2ax}dx=\frac{\sqrt{\pi}\sigma e^{2\sigma^2a^2}}{\sqrt{2}}\mathrm{erf}\left(\frac{x+2\sigma^2a}{\sqrt{2}\sigma}\right) \tag{11.32}$$

where $\mathrm{erf}(x)=\frac{2}{\sqrt{\pi}}\int_0^x e^{-t^2}dt$ is the error function. Therefore, using Eqs. (11.31) and (11.32), we can easily obtain the covariance matrix $\mathcal{R}_n$ of the expanded input vector.

$$\int e^{-\frac{x^2}{2\sigma^2}-2ax+2iB\pi x}dx=\frac{\sqrt{\pi}\sigma e^{2\sigma^2(a^2-2\pi aBi-\pi^2B^2)}}{\sqrt{2}}\mathrm{erf}\left(\frac{\sqrt{2}x+2\sqrt{2}\sigma^2(a-B\pi i)}{2\sigma}\right) \tag{11.33}$$

$$\int e^{-\frac{x^2}{2\sigma^2}-2ax-2iB\pi x}dx=\frac{\sqrt{\pi}\sigma e^{2\sigma^2(a^2+2\pi aBi-\pi^2B^2)}}{\sqrt{2}}\mathrm{erf}\left(\frac{\sqrt{2}x+2\sqrt{2}\sigma^2(a+B\pi i)}{2\sigma}\right) \tag{11.34}$$

In addition, we use the Taylor series expansion to compute the expectation of $\mathcal{G}(m)$ by approximating the mathematical expectation of its element $g(e_n(m))$ as

$$\begin{aligned} E\big[g(e_n(m))\big] &= E\left[\sum_{k=1}^{2} \frac{\lambda_k}{\sigma_k^2} \exp\left(-\frac{e_n^2(m)}{2\sigma_k^2}\right)\right] \\ &\approx E\left[\frac{\lambda_1}{\sigma_1^2}\left(1-\frac{e_n^2(m)}{2\sigma_1^2}\right)+\frac{\lambda_2}{\sigma_2^2}\left(1-\frac{e_n^2(m)}{2\sigma_2^2}\right)\right] \end{aligned} \tag{11.35}$$

When the step size η_n is small enough, we can assume that $E\big[e_n^2(m)\big] \approx \sigma_{v,n}^2$ holds. Thus, Eq. (11.35) can be rewritten as

$$E\big[g(e_n(m))\big] \approx \frac{\lambda_1}{\sigma_1^2}\left(1-\frac{\sigma_{v,n}^2}{2\sigma_1^2}\right)+\frac{\lambda_2}{\sigma_2^2}\left(1-\frac{\sigma_{v,n}^2}{2\sigma_2^2}\right) \tag{11.36}$$

After obtaining the expectations of $\mathcal{G}(m)$ and $\mathcal{Q}(m)$, we can obtain the convergence conditions of the proposed algorithm. Taking the mathematical expectation on both sides of Eq. (11.22) and using Assumptions A1–A4, we have

$$E\big[\tilde{\mathbf{h}}_g(m+1)\big] = E[\mathcal{B}(m)]E\big[\tilde{\mathbf{h}}_g(m)\big] \tag{11.37}$$

where $E[\mathcal{B}(m)] = \mathcal{A}(\mathbf{I}_{NP} - \mathcal{M}E[\mathcal{G}(m)]\mathcal{R})$, and $\mathcal{R} = E[\mathcal{Q}(m)] =$ blockdiag$\{\mathcal{R}_1, \mathcal{R}_2, \ldots, \mathcal{R}_N\}$ $\{\mathcal{R}_1, \mathcal{R}_2, \ldots, \mathcal{R}_N\}$.

From Eq. (11.37), it can be seen that the following condition needs to be satisfied if the algorithm is to converge,

$$\|E[\mathcal{B}(m)]\|_{b,\infty} < 1 \tag{11.38}$$

where $\|\cdot\|_{b,\infty}$ denotes the block maximum norm [38]. According to the properties of the block maximum norm, we have

$$\begin{aligned} \|E[\mathcal{B}(m)]\|_{b,\infty} &= \|\mathcal{A}(\mathbf{I}_{NP} - \mathcal{M}E[\mathcal{G}(m)]\mathcal{R})\|_{b,\infty} \\ &\leq \|\mathcal{A}\|_{b,\infty}\|\mathbf{I}_{NP} - \mathcal{M}E[\mathcal{G}(m)]\mathcal{R}\|_{b,\infty} \end{aligned} \tag{11.39}$$

Since $\mathcal{A} = \mathbf{A}^{\mathrm{T}} \otimes \mathbf{I}_P$, $\|\mathcal{A}\|_{b,\infty} = 1$ holds. Therefore, a sufficient condition for the algorithm to converge in the mean sense is written as

$$\|\mathbf{I}_{\mathrm{NP}} - \mathcal{M}E[\mathcal{G}(m)]\mathcal{R}\|_{b,\infty} = \max_{1\leq n\leq N}\left\|\mathbf{I}_P - \eta_n E\big[g(e_n(m))\big]\mathcal{R}_n\right\|_{b,\infty} \leq 1 \tag{11.40}$$

Consequently, a sufficient condition for the convergence of the proposed algorithm can be easily obtained, i.e., if η_n satisfies the following condition:

$$0 \leq \eta_n \leq \frac{2}{E\left[g(e_n(m))\right]\lambda_{\max}(\mathcal{R}_n)}, n = 1, 2, \ldots, N \tag{11.41}$$

where $\lambda_{\max}(\cdot)$ is the maximum eigenvalue of $\mathcal{R}_n$.

11.4.2 Transient Performance Analysis

In this subsection, we analyze the behavior of the mean square deviation (MSD) and excess mean square error (EMSE) of the NGD-MMCC algorithm. The MSD and EMSE at the network level are defined as

$$\mathrm{MSD}(m) = \frac{1}{N}E\left[\left\|\tilde{\mathbf{h}}_g(m)\right\|^2\right] \tag{11.42}$$

$$\mathrm{EMSE}(m) = \frac{1}{N}E\left[\left\|\tilde{\mathbf{h}}_g(m)\right\|^2_{\mathcal{R}}\right] \tag{11.43}$$

where $\left\|\tilde{\mathbf{h}}_g(m)\right\|^2_{\mathcal{R}} = \tilde{\mathbf{h}}_g^{\mathrm{T}}(m)\mathcal{R}\tilde{\mathbf{h}}_g(m)$.By taking the weighted square norm and the mathematical expectations on both sides of Eq. (11.22), we obtain

$$E\left[\left\|\tilde{\mathbf{h}}_g(m+1)\right\|^2_{\boldsymbol{\Sigma}}\right] = E\left[\left\|\tilde{\mathbf{h}}_g(m)\right\|^2_{\boldsymbol{\Sigma}'}\right] + E\left[\mathbf{c}^{\mathrm{T}}(m)\mathcal{A}^{\mathrm{T}}\mathcal{G}^{\mathrm{T}}(m)\mathcal{M}^{\mathrm{T}}\boldsymbol{\Sigma}\mathcal{M}\mathcal{G}(m)\mathcal{A}\mathbf{c}(m)\right] \tag{11.44}$$

where $\Sigma' = \mathcal{B}^{\mathrm{T}}(m)\boldsymbol{\Sigma}\mathcal{B}(m)$. To get a more compact and clearer form, a vectorization method is introduced, namely,

$$\sigma = \mathrm{vec}(\boldsymbol{\Sigma}) \tag{11.45}$$

$$\sigma' = \mathrm{vec}\left(\Sigma'\right) = \mathcal{F}\sigma \tag{11.46}$$

where $\mathcal{F} \approx E[\mathcal{B}(m)] \otimes E[\mathcal{B}(m)]$ [39]. The second term on the right-hand side of Eq. (11.44) can be written as

$$\begin{aligned}
&E\left[\mathbf{c}^{\mathrm{T}}(m)\mathcal{A}^{\mathrm{T}}\mathcal{G}^{\mathrm{T}}(m)\mathcal{M}^{\mathrm{T}}\sum\mathcal{M}\mathcal{G}(m)\mathcal{A}\mathbf{c}(m)\right] \\
&= \mathrm{Tr}\left(\mathcal{A}^{\mathrm{T}}\mathcal{M}^{\mathrm{T}}\sum\mathcal{M}\mathcal{A}E\left[\mathcal{G}^{\mathrm{T}}(m)\mathcal{G}(m)\right]\mathcal{S}\right) \\
&\approx \mathrm{Tr}\left(\mathcal{A}^{\mathrm{T}}\mathcal{M}^{\mathrm{T}}\sum\mathcal{M}\mathcal{A}E[\mathcal{G}(m)]E[\mathcal{G}(m)]\mathcal{S}\right) \\
&= \mathrm{Tr}\left(\sum\mathcal{P}\mathcal{S}\mathcal{P}^{\mathrm{T}}\right),
\end{aligned} \tag{11.47}$$

where $\mathcal{P} = \mathcal{AME}[\mathcal{G}(m)]$, and $\mathcal{S} = E\left[\mathbf{c}(m)\mathbf{c}^{\mathrm{T}}(m)\right] = \text{bolckdiag}\left\{\sigma_{v,1}^{2}\mathcal{R}_1, \sigma_{v,2}^{2}\mathcal{R}_2, \ldots, \sigma_{v,N}^{2}\mathcal{R}_N\right\}\left\{\sigma_{v,1}^{2}\mathcal{R}_1, \sigma_{v,2}^{2}\mathcal{R}_2, \ldots, \sigma_{v,N}^{2}\mathcal{R}_N\right\}$. Similarly, vectorizing the above equation yields

$$\mathrm{Tr}\left(\boldsymbol{\Sigma}\mathcal{P}\mathcal{S}\mathcal{P}^{\mathrm{T}}\right) = \mathcal{T}^{\mathrm{T}}\boldsymbol{\sigma} \tag{11.48}$$

where $\mathcal{T} = \mathrm{vec}\left(\mathcal{P}\mathcal{S}^{\mathrm{T}}\mathcal{P}^{\mathrm{T}}\right) = (\mathcal{P} \otimes \mathcal{P})\mathrm{vec}\left(\mathcal{S}^{\mathrm{T}}\right)$. Therefore, Eq. (11.44) can be rewritten as

$$E\left[\left\|\tilde{\mathbf{h}}_g(m+1)\right\|_{\sigma}^{2}\right] = E\left[\left\|\tilde{\mathbf{h}}_g(m)\right\|_{\mathcal{F}\sigma}^{2}\right] + \mathcal{T}^{\mathrm{T}}\boldsymbol{\sigma} \tag{11.49}$$

Iterating Eq. (11.49) from $m = 0$ yields

$$E\left[\left\|\tilde{\mathbf{h}}_g(m+1)\right\|_{\sigma}^{2}\right] = E\left[\left\|\tilde{\mathbf{h}}_g(0)\right\|_{\mathcal{F}^{m+1}\sigma}^{2}\right] + \mathcal{T}^{\mathrm{T}}\sum_{i=0}^{m}\mathcal{F}^{i}\boldsymbol{\sigma} \tag{11.50}$$

where $\tilde{\mathbf{h}}_g(0)$ denotes the initial expansion coefficient error. The relationship between $E\left[\left\|\tilde{\mathbf{h}}_g(m+1)\right\|_{\sigma}^{2}\right]$ and $E\left[\left\|\tilde{\mathbf{h}}_g(m)\right\|_{\sigma}^{2}\right]$ is derived by subtracting Eq. (11.50) at moment m from Eq. (11.50) at moment $m+1$, that is,

$$E\left[\left\|\tilde{\mathbf{h}}_g(m+1)\right\|_{\sigma}^{2}\right] = E\left[\left\|\tilde{\mathbf{h}}_g(m)\right\|_{\sigma}^{2}\right] + \mathcal{T}^{\mathrm{T}}\mathcal{F}^{i}\boldsymbol{\sigma} + E\left[\left\|\tilde{\mathbf{h}}_g(0)\right\|_{\left(\mathcal{F}^{m+1}-\mathcal{F}^{m}\right)\sigma}^{2}\right] \tag{11.51}$$

The transient behavior of MSD or EMSE of the proposed algorithm can be easily obtained by Eq. (11.51) when $\boldsymbol{\sigma} = \frac{1}{N}\mathrm{vec}(\mathbf{I}_{\mathrm{NP}})$ or $\boldsymbol{\sigma} = \frac{1}{N}\mathrm{vec}(\mathcal{R})$.

11.4.3 Steady-State Performance Analysis

The proposed algorithm is assumed to be stable and to reach a steady state after iterations. When the algorithm reaches steady state, it satisfies

$$\lim_{m\to\infty} E\left[\left\|\tilde{\mathbf{h}}_g(m+1)\right\|_{\sigma}^{2}\right] = \lim_{m\to\infty} E\left[\left\|\tilde{\mathbf{h}}_g(m)\right\|_{\sigma}^{2}\right] \tag{11.52}$$

Substituting Eq. (11.52) into Eq. (11.49), the steady-state equation can be written as

$$\lim_{m\to\infty} E\left[\left\|\tilde{\mathbf{h}}_g(m)\right\|_{\left(\mathbf{I}_{N^2P^2}-\mathcal{F}\right)\sigma}^{2}\right] = \mathcal{T}^{\mathrm{T}}\boldsymbol{\sigma} \tag{11.53}$$

Solving Eq. (11.53), the steady-state MSD and EMSE of the network can be obtained by

$$\lim_{m\to\infty} E\left[\| \tilde{\mathbf{h}}_g(m) \|_{\sigma}^{2}\right] = \mathcal{T}^{\mathrm{T}}(\mathbf{I}_{N^2P^2} - \mathcal{F})^{-1}\sigma \tag{11.54}$$

Similarly, steady-state MSD and EMSE are obtained when $\sigma = \frac{1}{N}\mathrm{vec}(\mathbf{I}_{\mathrm{NP}})$ and $\sigma = \frac{1}{N}\mathrm{vec}(\mathcal{R})$.

11.4.4 Complexity Analysis

In this subsection, we investigate the computational complexity of different nonlinear expansions. The number of operations for different nonlinear expansions is summarized in Table 11.1, where D is the order of the Volterra expansion. In addition, the computational complexity of the graph diffusion trigonometric functional least mean square (GD-TFLMS) [27] and NGD-MMCC algorithms are compared. The numbers of multiplication, addition and exponentiation operations per iteration on a single node are tabulated in Table 11.2.

It is seen from Table 11.1 that the complexities of all five nonlinear expansion appear to increase with length of the nonlinear expansion. For Volterra filter, the computational complexity of this nonlinear expansion exhibits a squared relationship with the order of the Volterra expansion. Therefore, its computational complexity is higher than the other four nonlinear expansions. In terms of efficiency in producing

Table 11.1 Computational complexities of five nonlinear expansions

Expansion type	Multiplication	Addition	Exponentiation	Cosine	Absolute value	Square root
Triangular expansion	$2LB + 2B - 2$	0	0	$2B$	0	0
Exponential expansion	$4LB + 4B - 2$	0	$2B$	$2B$	$2B$	0
Volterra	$LD/2 + LD^2/2$	0	0	0	0	0
Cosine feature	$LM + M + 1$	$M + L$-1	0	M	0	1
Gaussian feature	$LM + 3\,M$	$2LM$-M	M	0	0	M

Table 11.2 Computational complexities of different algorithms

Algorithms	Multiplication	Addition	Exponentiation
GD-TFLMS	$2P + 1$	$2P$	0
NGD-MMCC	$2P + 16$	$2P + 1$	2

the same length of nonlinear expansion, cosine feature and Gaussian feature are slightly worse than the other three methods. Furthermore, it is easy to see from Table 11.2 that the proposed algorithm has a similar computational complexity as existing nonlinear graph adaptive filtering algorithm.

11.5 Computer Simulation Examples

In this section, we use simulated and real data to confirm the practical effectiveness of the proposed algorithm and the accuracy of the theoretical analysis. When the nodes n and l are connected, the entries $s_{n,l}$ of the graph shift operator $\mathbf{S}$ can be derived using the subsequent equation:

$$s_{n,l} = \exp\left(\frac{\text{dist}_{n,l}^2}{\overline{\text{dist}}^2}\right) \tag{11.55}$$

where $\text{dist}_{n,l}$ denotes the Euclidean distance between nodes n and l and $\overline{\text{dist}}$ denotes the average Euclidean distance between all connected nodes. In addition, whether nodes are connected depends on the distance between nodes, i.e., we choose K nodes close to node n as its neighbor nodes. This selection method provides excellent representation of the correlation between data from nodes close to each other.

11.5.1 Verification of Transient and Steady-State Models

First, we validate the theoretical model of the NGD-MMCC algorithm based on exponential expansion in a randomly connected graph with nodes $N = 10$ and $K = 3$ shown in Fig. 11.1.

The node signal $x_n(m)$ is a zero-mean Gaussian process whose variance is sampled from the uniform distribution $\mathcal{U}(1, 1.5)$. The length of the graph filter is set to 2.

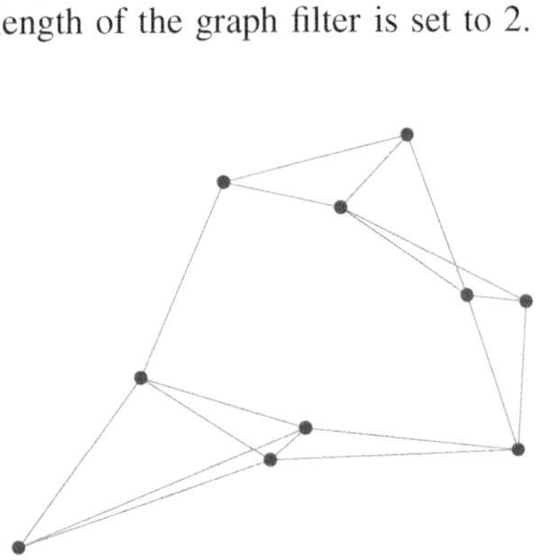

Fig. 11.1 Random graph topology with nodes $N = 10$

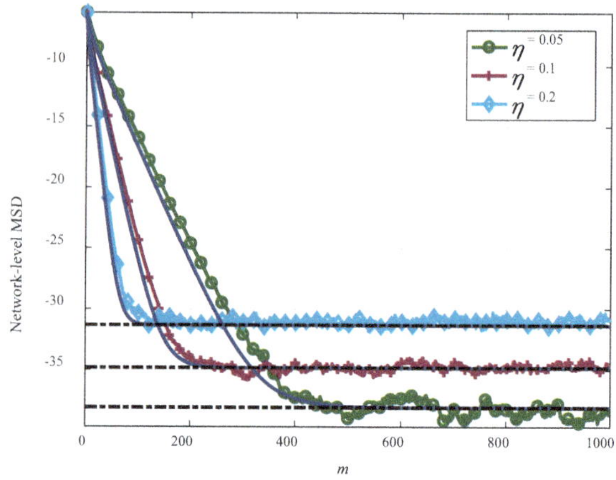

Fig. 11.2 Simulated and theoretical learning curves of the NGD-MMCC algorithm over different step sizes η

The noise $v_n(m)$ is considered for three different models that obey the Gaussian distribution $\mathcal{N}(0, 0.01)$ in Fig. 11.2, the uniform distribution $\mathcal{U}(-b, b)$ in Fig. 11.4 and the binary distribution over $\{-0.4, 0.6\}$ with probability $Pr(v_n(m) = -0.4) = 0.6$ in Fig. 11.6. The combination coefficient $a_{l,n}$ is calculated as $1/|\mathcal{N}_n|$, where $|\cdot|$ is the cardinality of the set. The noisy output $y_n(m)$ is obtained by a nonlinear LIP model $y_n(m) = \mathbf{q}_n^{\mathrm{T}}(m)\mathbf{h}_o + v_n(m)$. The optimal expansion coefficients $\mathbf{h}_o$ are generated randomly. In addition, all simulated curves are obtained by averaging over 50 independent Monte Carlo runs.

The theoretical and simulated learning curves of the NGD-MMCC algorithm for different step sizes, noise variances, kernel bandwidths, and mixture parameters are plotted in Figs. 11.2, 11.3, 11.4 and 11.5, respectively. The default parameter chosen are $a = 0.5$, $b = 0.4$, $\mu = 0.1$, $\sigma_1 = 0.8$, $\sigma_2 = 2$ and $\lambda_1 = 0.5$. The dotted line refers to the simulation result, and the solid and dashed lines represent the transient and steady-state results generated by (11.51) and (11.54), respectively. As shown in Figs. 11.2, 11.3, 11.4 and 11.5, the theoretical curve matches very well with the simulation results. The steady-state MSD of the NGD-MMCC algorithm with respect to step size η_n is given in Fig. 11.6. As can be seen in Fig. 11.6, the upper bound of the step size based on Eq. (11.41) is consistent with the simulation results. The simulation results demonstrate the accuracy and correctness of our derived model.

11.5.2 Nonlinear System Identification

Second, two typical nonlinear system models are used to compare the performance of the proposed algorithm in a non-Gaussian noise environment. The graph topology with $N = 20$ and $K = 3$ is given in Fig. 11.7. The noise $v_n(m)$ is sampled from multiplicative model, defined as $v_n(m) = b_n(m)A_n(m)$, where $b_n(m)$ obeys a binary distribution over $\{0, 1\}$ with the probability mass function $Pr(b_n(m) = 1) = 0.2$,

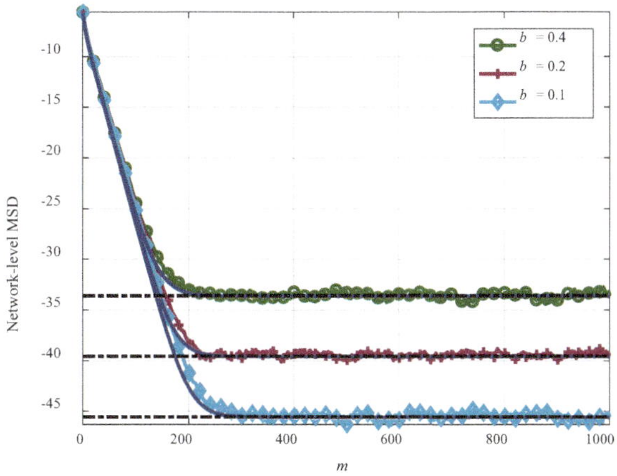

Fig. 11.3 Simulated and theoretical learning curves of the NGD-MMCC algorithm over different noise variances

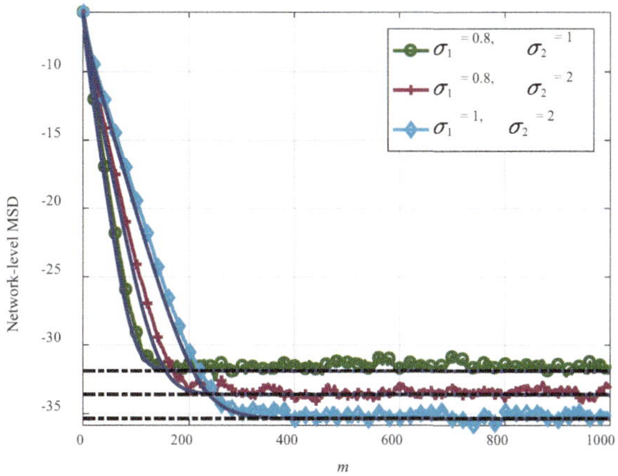

Fig. 11.4 Simulated and theoretical learning curves of the NGD-MMCC algorithm over different kernel bandwidths

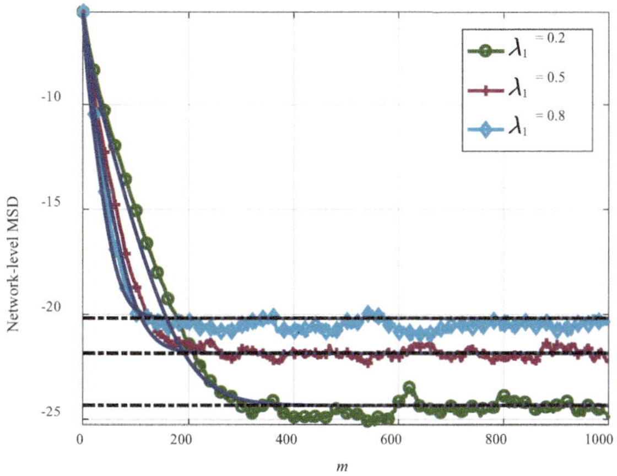

Fig. 11.5 Simulated and theoretical learning curves of the NGD-MMCC algorithm over different mixture parameters

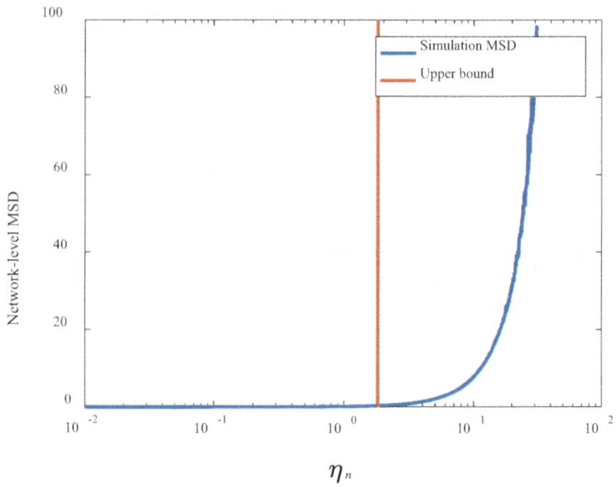

Fig. 11.6 Steady-state MSD of the NGD-MMCC algorithm versus step size η_n

and $A_n(m)$ is an alpha-stable noise with the tail parameter $\alpha = 1.2$ and dispersion parameter $\gamma = 0.1$. The length of the graph filter is 4, and the nonlinear system models are given by [25, 27].

$$d_n(m) = 0.6\sin^3\big(k_{n,1}(m)\big) - \frac{2}{k_{n,1}^3(m)+2} - 0.1\cos\big(4\pi k_{n,4}(m)\big) + 1.125 + v_n(m) \tag{11.56}$$

$$d_n(m) = \sqrt{k_{n,1}^2(m) + \sin^2\big(k_{n,4}(m)\pi\big)} + \big(0.8 - 0.5\exp\big(-k_{n,2}^2(m)\big)\big)k_{n,3}(m) + v_n(m) \tag{11.57}$$

In models (11.56) and (11.57), the graph signals are Gaussian processes with zero mean, and their variance are generated from the uniform distributions [0.3, 0.5] and [0.5, 1], respectively. We compared the proposed algorithms based on different nonlinear expansions with GD-TFLMS [27] and graph diffusion random Fourier

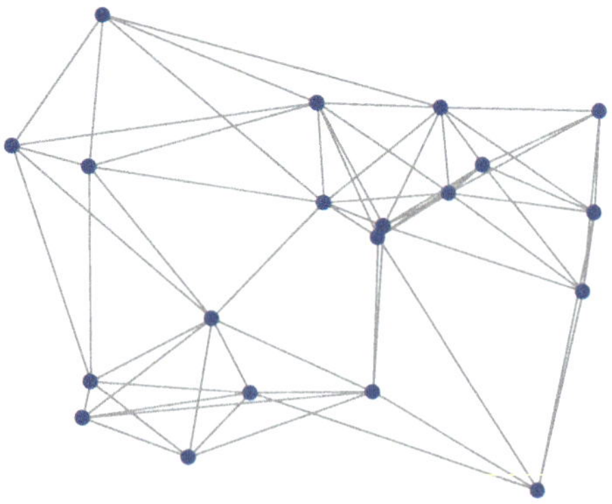

Fig. 11.7 Random graph topology with nodes $N = 20$

features least mean square (GD-RFFLMS) [25]. The simulation parameters used are: $\sigma_1 = 0.8$, $\sigma_2 = 2$, $\lambda_1 = 0.5$, $\lambda_2 = 0.5$, $B = 3$, $a = 0.8$, $M = 28$, $\sigma = 1$ and the order of the Volterra expansion is chosen to be 3. The simulation results for the different algorithms are shown in Figs. 11.8 and 11.9. As expected, the proposed algorithm can effectively deal with nonlinear identification problems in non-Gaussian environments compared to the nonlinear adaptive filtering algorithm based on the MMSE criterion. In addition, since different nonlinear models have different degrees of nonlinearity, it is necessary to employ various nonlinear expansions for different graph signals and system models to achieve optimal algorithm performance.

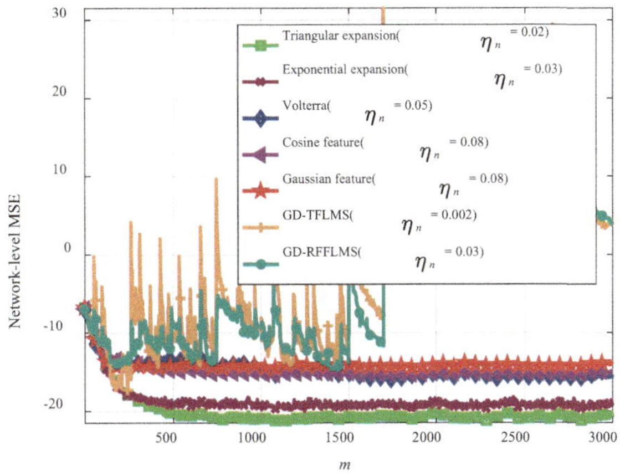

Fig. 11.8 Comparison of convergence curves of network-level MSE for different algorithms in model (11.56)

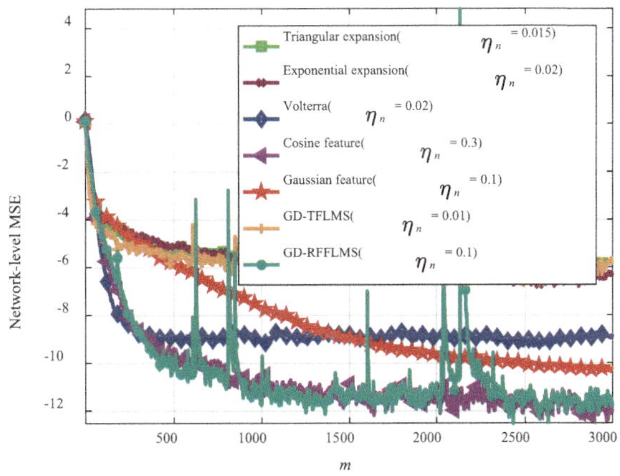

Fig. 11.9 Comparison of convergence curves of network-level MSE for different algorithms in model (11.57)

Fig. 11.10 Graph based on the 12th hour data in the U.S. dataset

11.5.3 Prediction in Temperature Sensor Networks

Third, we evaluate the tracking performance of the proposed algorithm using the temperature data measured in the first 95 h of January 2010 at 205 U.S. temperature stations. The length of the graph filter is 4. The graph structure with $K = 7$ of the 12th hour in the dataset is given in Fig. 11.10. Figure 11.11 shows the temperature signal of the circled nodes in Fig. 11.10 over time and compares the tracking capabilities of the NGD-MMCC algorithms based on different nonlinear expansions with diffusion MCC (DMCC) algorithm [40]. In the simulation, the parameters of the different algorithms are set as follows: trigonometric expansion $\eta_n = 0.1, B = 4$, exponential expansion $\eta_n = 0.1, B = 4, a = 0.2$, Volterra expansion $D = 2$, $\eta_n = 0.018$, cosine feature function $M = 56$, $\eta_n = 0.8, \sigma = 1$, Gaussian feature $M = 16$, $\eta_n = 1.8, \sigma = 3$, and DMCC $\eta_n = 0.001$, $\sigma = 1$. The parameters used in MMCC are $\sigma_1 = 1, \sigma_2 = 2, \lambda_1 = 0.5$ and $\lambda_2 = 0.5$. Table 11.3 gives a comparison of MSE accuracy for different algorithms. It can be seen from the Fig. 11.11 that the proposed algorithm is able to predict the node temperature better than the DMCC algorithm and can effectively handle temperature prediction for sensor networks. In addition, it is observed from Table 11.3 that the accuracy of the algorithms based on triangular, exponential and Volterra expansions is better than that of the algorithms based on cosine and Gaussian features.

11.6 Summary

In this paper, to handle the nonlinear modeling problem in a non-Gaussian noise environment, an NGD-MMCC algorithm is proposed by combining the LIP model. In addition, the convergence conditions as well as the transient and steady-state behaviors of the proposed algorithm are analyzed from a statistical point of view. The simulation results show that the theoretical model matches well with the simulation

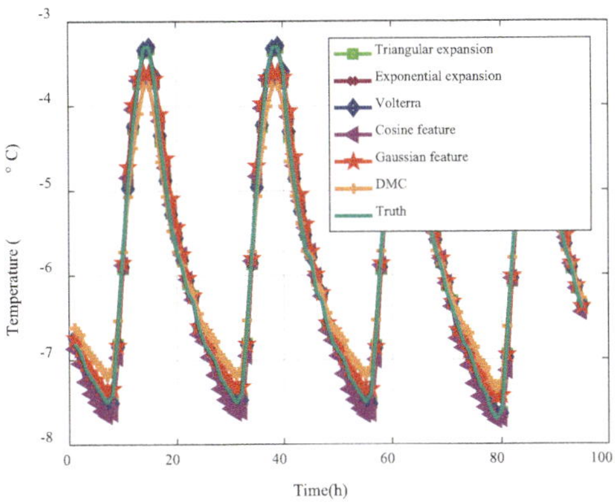

Fig. 11.11 Comparison of tracking capabilities of the NGD-MMCC algorithms based on various nonlinear expansions with DMCC algorithm

Table 11.3 Accuracy of different algorithms

Algorithms	MSE/dB
Triangular expansion	−21.4855
Exponential expansion	−21.4854
Volterra	−21.4848
Cosine feature	−15.0297
Gaussian feature	−15.3358
DMCC	−11.8540

results and the proposed algorithms exhibit strong robustness in the non-Gaussian noise environment. It is worth noting that the algorithms based on different nonlinear expansions show relatively large discrepancies in different system models and input signals. Therefore, how to choose a suitable nonlinear expansion is still an open problem, which will be one of our future research interests.

References

1. I. Jablonski, Graph signal processing in applications to sensor networks, smart grids, and smart cities. IEEE Sensors J. **17**(23), 7659–7666 (2017)
2. A.C. Yagan, M.T. Özgen, Spectral graph based vertex-frequency Wiener filtering for image and graph signal denoising. IEEE Trans. Signal Inf. Process. Netw. **6**, 226–240 (2020)
3. T. Wang, H. Guo, X. Yan, Z. Yang, Speech signal processing on graphs: the graph frequency analysis and an improved graph Wiener filtering method. Speech Commun. **127**, 82–91 (2021)
4. L. Wan, X. Li, J. Xu, L. Sun, X. Wang, K. Liu, Application of graph learning with multivariate relational representation matrix in vehicular social networks. IEEE Trans. Intell. Transp. Syst. **24**(3), 2789–2799 (2023)
5. L. Stankovic, M. Daković, E. Sejdić, *Introduction to Graph Signal Processing (Communications and Signal Processing)* (Cambridge University Press, Cambridge, U.K., 2018)
6. M. Barthelemy, *Morphogenesis of Spatial Networks* (Springer, Cham, Switzerland, 2018)
7. A.H. Sayed, *Fundamentals of Adaptive Filtering* (Wiley, Hoboken, NJ, USA, 2003)

8. H. Zhao, X. Zeng, X. Zhang, Z. He, T. Li, W. Zhao, Adaptive extended pipelined second-order Volterra filter for nonlinear active noise controller. IEEE/ACM Trans. Audio Speech Lang. Process. **20**(4), 1394–1399 (2012)
9. H. Zhao, J. Zhang, A novel adaptive nonlinear filter-based pipelined feedforward second-order Volterra architecture. IEEE Trans. Signal Process. **57**(1), 237–246 (2009)
10. H. Zhao, J. Zhang, Pipelined Chebyshev functional link artificial recurrent neural network for nonlinear adaptive filter. IEEE Trans. Syst. Man Cybern. Syst. **40**(1), 162–172 (2010)
11. D. Comminiello, M. Scarpiniti, L.A. Azpicueta-Ruiz, J. Arenas-García, A. Uncini, Functional link adaptive filters for nonlinear acoustic echo cancellation. IEEE/ACM Trans. Audio Speech Lang. Process. **21**(7), 1502–1512 (2013)
12. S. Zhang, W.X. Zheng, Recursive adaptive sparse exponential functional link neural network for nonlinear AEC in impulsive noise environment. IEEE Trans. Neural Netw. Learn. Syst. **29**(9), 4314–4323 (2018)
13. B.D. Chen, S.L. Zhao, P.P. Zhu, J.C. Principe, Quantized kernel recursive least squares algorithm. IEEE Trans. Neural Netw. Learn. Syst. **24**(9), 1484–1491 (2013)
14. Y. Engel, S. Mannor, R. Meir, The kernel recursive least-squares algorithm. IEEE Trans. Signal Process. **52**(8), 2275–2285 (2004)
15. W.D. Parreira, J.C.M. Bermudez, C. Richard, J.-Y. Tourneret, Stochastic behavior analysis of the Gaussian kernel least-mean-square algorithm. IEEE Trans. Signal Process. **60**(5), 2208–2222 (2012)
16. H. Zhao, W. Xiang, S. Lv, A variable parameter LMS algorithm based on generalized maximum correntropy criterion for graph signal processing. IEEE Trans. Signal Inf. Process. Netw. **9**, 140–151 (2023)
17. S. Lv, H. Zhao, W. Xu, Robust widely linear affine projection M-estimate adaptive algorithm: Performance analysis and application. IEEE Trans. Signal Process. **71**, 3623–3636 (2023)
18. H. Zhao, G. Wang, F. Zhao, D. Liu, P. Song, Recursive general mixed norm algorithm for censored regression: performance analysis and channel equalization application. IEEE Trans. Syst. Man Cybern. Syst. **54**(2), 752–763 (2024)
19. J.C. Principe, *Information Theoretic Learning: Renyi's Entropy and Kernel Perspectives* (Springer Verlag, Berlin, Germany, 2010)
20. W. Liu, P.P. Pokharel, J.C. Principe, Correntropy: properties and applications in non-Gaussian signal processing. IEEE Trans. Signal Process. **55**(11), 5286–5298 (2007)
21. H. Zhao, D. Liu, S. Lv, Robust maximum correntropy criterion subband adaptive filter algorithm for impulsive noise and noisy input. IEEE Trans. Circ. Syst. II Express Briefs **69**(2), 604–608 (2022)
22. Y. Zheng, B. Chen, S. Wang, W. Wang, W. Qin, Mixture correntropy-based kernel extreme learning machines. IEEE Trans. Neural Netw. Learn. Syst. **33**(2), 811–825 (2022)
23. B. Chen, X. Wang, N. Lu, S. Wang, J. Cao, J. Qin, Mixture correntropy for robust learning. Pattern Recogn. **79**, 318–327 (2018)
24. P. Di Lorenzo, S. Barbarossa, P. Banelli, S. Sardellitti, Adaptive least mean squares estimation of graph signals. IEEE Trans. Signal Inf. Process. Netw. **2**(4), 555–568 (2016)
25. V.R.M. Elias, V.C. Gogineni, W.A. Martins, S. Werner, Adaptive graph filters in reproducing kernel Hilbert spaces: design and performance analysis. IEEE Trans. Signal Inf. Process. Netw. **7**, 62–74 (2021)
26. V. C. Gogineni, V. Naumova, S. Werner, Y. Huang, Graph kernel recursive least-squares algorithms, in *Asia-Pacific Signal and Information Processing Association Annual Summit and Conference (APSIPA ASC)* (2021), pp. 2072–2076
27. L. Li, Y.-F. Pu, Z.-Y. Luo, Distributed functional link adaptive filtering for nonlinear graph signal processing. Dig. Signal Process. **128**, 103558 (2022)
28. Z. Xiao, X. Wang, Nonlinear polynomial graph filter for signal processing with irregular structures. IEEE Trans. Signal Process. **66**(23), 6241–6251 (2018)
29. X. Hou, H. Zhao, X. Long, Graph diffusion kernel maximum correntropy criterion over sensor network and its performance analysis. IEEE Sens. J. **23**(13), 14583–14591 (2023)

30. A. Carini, G.L. Sicuranza, V.J. Mathews, Efficient adaptive identification of linear-in-the-parameters nonlinear filters using periodic input sequences. Signal Process. **93**(5), 1210–1220 (2013)
31. S.S. Bhattacharjee, N.V. George, Nearest Kronecker product decomposition based linear-in-the-parameters nonlinear filters. IEEE/ACM Trans. Audio Speech Lang. Process. **29**, 2111–2122 (2021)
32. Y. Zhu, H. Zhao, X. He, Z. Shu, B. Chen, Cascaded random Fourier filter for robust nonlinear active noise control. IEEE/ACM Trans. Audio Speech Lang. Process. **30**, 2188–2200 (2022)
33. J. Dong, Y. Zheng, B. Chen, A unified framework of random feature KLMS algorithms and convergence analysis, in *International Joint Conference on Neural Networks* (2018), pp. 1–8
34. F.S. Cattivelli, A.H. Sayed, Diffusion LMS strategies for distributed estimation. IEEE Trans. Signal Process. **58**(3), 1035–1048 (2010)
35. L. Li, H. Zhao, A robust total least mean M-estimate adaptive algorithm for impulsive noise suppression. IEEE Trans. Circ. Syst. II Express Briefs **67**(4), 800–804 (2020)
36. X. Hou, H. Zhao, X. Long, An innovative transient analysis of adaptive filter with maximum correntropy criterion. IEEE Signal Process. Lett. **29**, 1689–1693 (2022)
37. E. Cho, De Moivre's formula for quaternions. Appl. Math. Lett. **11**(6), 33–35 (1998)
38. A. Sayed, Adaptation, learning, and optimization over networks. Found. Trends Mach. Learn. **7**(4–5), 311–801 (2014)
39. A. H. Sayed, Diffusion adaptation over networks, in *Academic Press Library in Signal Processing*, vol. 3 (O'Reilly, Boston, MA, USA, 2014), pp. 323–453
40. W. Ma, B. Chen, J. Duan, H. Zhao, Diffusion maximum correntropy criterion algorithms for robust distributed estimation. Dig. Signal Process. **58**, 10–19 (2016)

Index

H. Zhao et al., *Information Theoretic Learning-Based Filter*,
https://doi.org/10.1007/978-3-032-29623-8

GPSR Compliance

The European Union's (EU) General Product Safety Regulation (GPSR) is a set of rules that requires consumer products to be safe and our obligations to ensure this.

If you have any concerns about our products, you can contact us on ProductSafety@springernature.com

In case Publisher is established outside the EU, the EU authorized representative is:

Springer Nature Customer Service Center GmbH
Europaplatz 3
69115 Heidelberg, Germany

Batch number: 10784019

Printed by Printforce, the Netherlands